Procedures to Investigate Waterborne Illness

Procedures to Investigate Waterborne Illness

Third Edition

Prepared by the
Committee on the Control of Foodborne Illness
International Association for Food Protection

Marilyn B. Lee and Ewen C. D. Todd. 2016. *Procedures to Investigate Waterborne Illness*, 3rd edition, International Association for Food Protection, Des Moines, Iowa.

ISBN 978-3-319-26025-9   ISBN 978-3-319-26027-3   (eBook)
DOI 10.1007/978-3-319-26027-3

Library of Congress Control Number: 2016931612

Printed on acid-free paper

This Springer imprint is published by Springer Nature
The registered company is Springer International Publishing AG Switzerland

Procedures to Investigate Waterborne Illness

Third Edition

Prepared by the
Committee on the Control of Foodborne Illness
International Association for Food Protection

The following individuals contributed to this edition:

Michael Brodsky, M.Sc.
Brodsky Consultants
Thornhill, ON, Canada

Nelson Fok, M.Sc., C.P.H.I. (c)
Department of Public Health
Concordia University of Edmonton
Edmonton, AB, Canada

Jack Guzewich*, M.P.H., R.S., F.D.A. C.F.S.A.N. (Retired)
Albany, NY, USA

Julie Ann Kase, Ph.D.
US Food and Drug Administration/CFSAN
College Park, MD, USA

Marilyn B. Lee*, Sc.M., C.P.H.I. (c)
School of Occupational and Public Health
Ryerson University
Toronto, ON, Canada

Carrie Lewis, M.Sc.
Milwaukee Water Works
Milwaukee, WI, USA

Cassandra D. LoFranco, M.Sc.
Ontario Scientific Advisors Inc.
Georgetown, ON, Canada

Sharon Nappier, Ph.D., M.S.P.H.
Office of Water, Office of Science and Technology
Health and Ecological Criteria Division
Human Health Risk Assessment Branch
US Environmental Protection Agency (4304T)
Washington, DC, USA

Per Nilsson*, B.S.
Profox Company
Kristdala, Sweden

Timothy Sly, Ph.D., C.P.H.I. (c)
School of Occupational and Public Health
Ryerson University
Toronto, ON, Canada

Ewen C.D. Todd*, Ph.D.
Committee Chair
Ewen Todd Consulting
Okemos, MI, USA and
Large Animal Clinical Sciences
Michigan State University
Lansing, MI, USA

*Indicates those who are Committee Members

Foreword

Procedures to Investigate Waterborne Illness is designed to guide public health, environmental protection, engineering, and other personnel who investigate reports of illnesses alleged to be waterborne. Prompt action to control an outbreak at its start will help minimize illness. This manual is an update of the 1996 second edition of *Procedures to Investigate Waterborne Illness*. The text has been adjusted where appropriate to reflect more recent understanding of pathogens and their control. A section on the role of environmental water, for instance from irrigation and processing of produce, has been added and the "Collection and Analysis of Data" section has been significantly expanded. The manual is a companion document to the *Procedures to Investigate Foodborne Illness* which was revised and published in 2012. Both the foodborne and waterborne illness manuals are based on epidemiologic principles and investigative techniques that have been found effective to determining causal factors of disease outbreaks. They are designed to improve the quality of investigation of outbreaks and disease surveillance. The reader will note that there is much overlap between the two manuals as the gathering of information during an investigation is very similar as are many of the pathogens that are capable of causing either foodborne or waterborne illness. In fact at the start of the investigation it will often not be clear whether an outbreak is foodborne or waterborne. There are differences in that whereas most waterborne illnesses are contracted through ingestion similar to food, infections can also be generated through skin contact and aerosols.

The *Table of Contents* serves as an outline and a flow diagram shows the interrelationships of the activities and their typical sequence in an investigation. The topics in the manual are presented in the sequence usually followed during investigations, i.e., receiving complaints of an alleged outbreak, interviewing the ill (and sometimes the healthy), developing a case definition, collecting water samples and clinical samples, conducting investigations of the water source and its distribution, interpreting the data, and reporting the outbreak. They are organized so that an investigator can easily find the information needed at any phase of the investigation. The forms at the end offer a convenient and effective way to keep information organized and ready for analysis. And the tables are useful in providing more detailed information during an investigation, although they are not exhaustive. The tables on the pathogens are organized by symptoms, which will often be known early in an investigation before the pathogen has been confirmed by the laboratory.

Although this manual is intended to aid the investigator when an outbreak happens, the information in part could also be used as a guideline in prevention of an outbreak. Having adequate oversight of a water system and having trained staff are key to preventing outbreaks.

Contents

Procedures to Investigate Waterborne Illness

Introduction

Humanity could not survive without a reliably clean, safe, and steady flow of drinking water. Since the early 1900s when typhoid fever and cholera were frequently causes of waterborne illness in developed countries, drinking water supplies have been protected and treated to ensure water safety, quality, and quantity. Having access to safe drinking water has always been one of the cornerstones of good public health. Safe water is not limited to drinking water, since recreational water and aerosolized water can also be sources for waterborne illness, from treated waters such as in swimming pools, whirlpools, or splash pads and from non-treated surface waters such as lakes, rivers, streams and ponds. Recreational waters may cause illness not only from ingestion of pathogens, but also when in contact with eyes, ears, or skin. Some pathogens in water can be acquired by inhalation of aerosols from water that is agitated or sprayed such as in humidifiers, fountains, or misting of produce. This poses a potential risk to those exposed, particularly if they are immunocompromised.

Often when an outbreak is first suspected, the source is not clear, i.e., food, water, animal contact. Investigation is usually needed to find the common source. In some outbreaks the food may first be identified as the source, such as with produce, but the ultimate source could be contaminated irrigation water. Investigators have to keep an open mind until laboratory and/or epidemiologic evidence links cases to the primary source.

Although we frequently think of waterborne illness originating from a microbiological agent, we should be aware that water may also be contaminated by pesticides, fertilizers, and other chemicals which may enter through industrial discharge, agriculture runoff, or deliberate contamination.

© International Association for Food Protection 2016
Food Protection International Association, *Procedures to Investigate
Waterborne Illness*, DOI 10.1007/978-3-319-26027-3_1

1

Waterborne illness acquired from microorganisms may be classified as:

- Toxin-mediated infections caused by bacteria that produce enterotoxins or emetic toxins that affect water, glucose, and electrolyte transfer during their colonization and growth in the intestinal tract;
- Infections caused when microorganisms invade and multiply in the intestinal mucosa, eyes, ears, or respiratory tract, or contact the skin;
- Intoxications caused by ingestion of water containing poisonous chemicals or toxins produced by other microorganisms

Manifestations range from slight discomfort to acute illness to severe reactions that may terminate in death or chronic sequelae, depending on the nature of the causative agent, number of pathogenic microorganism or concentration of poisonous substances ingested, and host susceptibility and reaction.

The public relies on public health regulators to investigate and mitigate waterborne illness. Mitigation depends upon rapid detection of outbreaks and a thorough knowledge of the agents and factors responsible for waterborne illness. Public health and law enforcement agency officials should always be alert to the rare possibility of an intentional contamination of water supplies by disgruntled employees or terrorists.

The purposes of a waterborne illness investigation are to stop the outbreak or prevent further exposure by:

- Identifying illness associated with an exposure and verifying that the causative agent is waterborne
- Detecting all cases, the causative agent, and the place of exposure
- Determining the water source, mode of contamination, processes, or practices by which proliferation and/or survival of the etiological agent occurred
- Implementing emergency measures to control the spread of the outbreak
- Gathering information on the epidemiology of waterborne diseases and the etiology of the causative agents that can be used for education, training, and program planning, thereby impacting on the prevention of waterborne illness
- Determining if the outbreak under investigation is part of a larger outbreak by immediately reporting to state/provincial/national epidemiologists

In the instance of a bottled water outbreak, halting of distribution and sale of product and recall of product, some of which may already be in consumers' homes, are necessary to prevent further illness.

As epidemiologic data accumulate, information will indicate the source of the problem, whether a municipal water treatment plant, bottled water manufacturing plant, or recreational water exposure, and suggest methods for controlling and preventing waterborne illness. This information will guide administrators in making informed decisions to provide the highest degree of waterborne safety.

A flowchart, *Sequence of events in investigating a typical outbreak of waterborne illness* (Fig. 1) shows the sequential steps, as presented in this manual, in investigating a typical outbreak of waterborne illness and illustrates their relationships. A description of each step is presented in this manual.

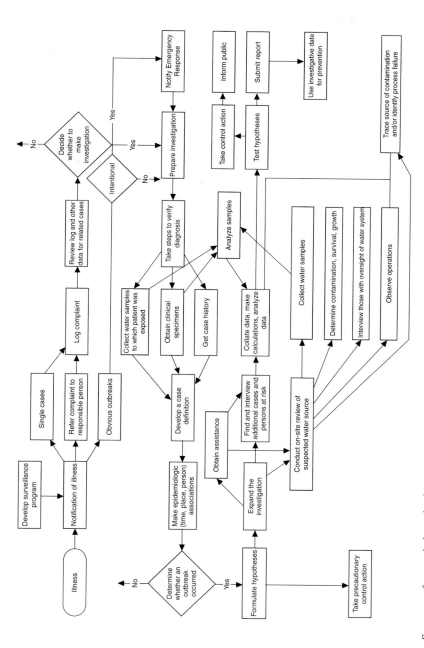

Fig. 1 Sequence of events in investigating a typical outbreak of waterborne illness. An intentional water contamination event may or may not be obvious. It can be recognized at any point during the outbreak investigation. If intentional contamination is suspected follow your notification scheme in emergency response plans (this could include law enforcement, emergency management and other government agencies)

Develop a Waterborne Disease Surveillance and Emergency Operations Program

The primary purpose of a waterborne disease surveillance system is to systematically gather accurate information on the occurrence of water-related illnesses in a community, thus allowing development of a rational approach for the detection, control and prevention of waterborne illness. Other purposes are to (a) determine trends in the incidence of waterborne diseases, (b) characterize the epidemiology of waterborne diseases, (c) gather and disseminate information on waterborne diseases, and (d) develop a basis for evaluating control efforts. It may be useful to coordinate this system with, or integrate it into a foodborne disease surveillance system. However, while the procedures are quite similar from an epidemiologic viewpoint, they may differ with respect to personnel or agencies involved. An effective disease surveillance system is essential for detection of disease caused by either unintentional or intentional contamination of food.

An effective waterborne disease surveillance system consists of:

- Early reports of enteric and other illnesses that may be related to water exposure or consumption
- Coordinated effort between local and state public health partners, water utility and water recreation staff
- Systematic organization and interpretation of data
- Timely investigation of identified outbreaks or clusters of illness
- Dissemination of outbreak reports and surveillance summaries to all appropriate stakeholders

Many types of reporting systems may already exist at the local or state/provincial level, and these should be incorporated into a waterborne disease surveillance program. These include (a) mandatory (or voluntary) laboratory- or physician-based reporting of specific infectious diseases, (b) national-based surveillance systems such as CaliciNet (CDC 2009) or NORS (CDC 2009) in the US, (c) physician office, hospital emergency, and urgent-care clinic medical records, (d) public complaints made to health agencies and/or local water utilities, (e) school illness and absentee records, (f) absentee records of major employers, (g) water treatment records kept by water utilities (e.g., turbidity, disinfection levels, occurrence of coliforms), (h) increased sales of antidiarrheal drugs and anti-nausea medications, and (i) source water quality data kept by environmental agencies (e.g., departments of natural resources and geological survey agencies). Another type of surveillance mechanism that may supplement or enhance existing reporting systems is a daily log of illness and water quality complaints.

Organize the System and Develop Procedures

An effective waterborne illness surveillance system requires close cooperation between key personnel in public and private health agencies, laboratories, water utilities and water recreation staff, and environmental health agencies. When your agency contemplates initiation or development of a waterborne illness surveillance program, give top priority to identification of appropriate financial, political, strategic, and administrative support. Then, identify a key person to create, implement, and manage the system.

This person takes responsibility for:

- Reviewing the types of reporting systems that already exist in your agency or in other agencies that could be incorporated into a waterborne illness surveillance system
- Identifying the types of information that cannot be obtained from existing reporting systems but that need to be collected or addressed by the waterborne illness surveillance system
- Identifying ways to merge or integrate the data collected by existing systems with data gathered in the waterborne illness surveillance system
- Identifying collaborating agencies and staff
- Develop a mechanism to communicate and update all stakeholders (may be by blast e-mail or periodic conference calls)
- Providing training in surveillance methods for agency staff and other partners to enhance cooperation
- Assembling materials that will be required during an outbreak investigation
- Evaluating the effectiveness of the system.

Develop procedures to seek and record complaints about waterborne illnesses, water supplies, and water recreational sites. For example, list the telephone number of the waterborne illness investigation unit prominently on local and state public health and water utility websites. To be most effective, have this number monitored 24/7 by staff or an answering service. If possible, the utilization of social media such as Facebook or Twitter should be considered and monitored as many large municipalities (including drinking water utilities) and recreational facilities have an Internet presence. If your agency has social media accounts, consider using this vehicle to further disseminate information regarding waterborne illness clusters or outbreaks. Identify medical care facilities and practitioners and seek their participation. Direct educational activities, such as newsletters and talks at meetings, to stimulate participation in the program. Encourage water treatment utilities and operators of recreational water sites to report suspected complaints of waterborne illness to the appropriate local agencies. Also, encourage private and hospital laboratories to report isolations of parasitic agents (e.g., *Giardia, Cryptosporidium*), viruses (e.g., norovirus and hepatitis A virus), bacteria (e.g., *E. coli* (pathogenic), *Salmonella, Shigella, Vibrio cholerae*), and other agents that may be waterborne. Develop a protocol for notification and coordination with agencies that might cooperate in investigational activities, including 24-h-a-day, 7-days-a-week contacts. A comprehensive contact list should be constructed and updated at least twice a year as individuals may change. Notify and coordinate with state/provincial or district agencies, national agencies that have surveillance and water regulatory responsibilities, and other national and international health agencies, as appropriate. For example, it may be useful to find out the level of participation within a certain jurisdiction in national-level outbreak surveillance programs such as NORS (CDC, 2015) or other national surveillance system.

Assign Responsibility

Delegate responsibility to a professionally trained person who is familiar with epidemiologic methods and with the principles of water treatment and recreational water protection. This person will (a) direct the surveillance program, (b) take charge if waterborne and enteric outbreaks are suspected, and (c) handle publicity during outbreaks. Delegate responsibility to others who will carry out specific epidemiologic, laboratory and on-site investigations. If an intentional contamination event is suspected, local and national law enforcement agencies will likely become the lead agency responsible for the investigation. With this in mind, it is critical to identify appropriate individuals and include them in communication and any practice drills that may occur. If a relationship has been established prior to any event, the investigation may run more smoothly.

Establish an Investigation Team

Enlist help from a team of epidemiologists, microbiologists, sanitarians/environmental health officers/public health inspectors, engineers, chemists, nurses, physicians, public information specialists, and other (e.g., toxicologists) as needed. Free flow of information and coordination among those participating in waterborne disease surveillance and investigation are essential, particularly when several different agencies are involved. Water-related complaints are equally likely to be directed at health departments or water utilities but also perhaps to different jurisdictions. Therefore, it is essential that these complaints be registered by an agency and that the information is rapidly shared within and perhaps outside of a particular jurisdiction as part of an integrated surveillance system. Whenever possible, share the information with participating parties by rapid means such as e-mail and by calling 24/7 contact phone numbers. If an intentional contamination event is suspected, contact emergency response and law enforcement for their early involvement.

Train Staff

Select staff members who will participate in the waterborne disease surveillance program on the basis of interest and ability. Inform them of the objectives and protocol of the program. Emphasize not only the value of disease surveillance, but also the value of monitoring water quality and treatment performance. If possible, provide printed learning material that can be referenced later. Encourage the use of epidemiologic information and approaches in routine disease surveillance and prevention activities. Develop their skills so that they can carry out their role effectively during an investigation, and teach them how to interpret data collected during investigations. Conduct seminars routinely and during or after investigations to update staff and keep agency personnel informed. Train office workers who will receive calls concerning waterborne illnesses to give appropriate instructions. Those who participate in the investigation will learn from the experience and often are in a position to implement improvements after completion of the investigation.

Assemble Materials

Assemble and have readily available kits with forms and equipment as specified in Table A (Equipment useful for investigations). Restock and maintain kits on a schedule recommended by, and in cooperation with, laboratory staff to ensure their stability and sterility. Verify expiration dates, and use kits before this date or discard and reorder. Assemble a reference library on waterborne illnesses, investigation techniques, and control measures from reference books, scientific journal articles, manuals, and reputable Internet sources (e.g., www.cdc.gov, www.who.int/en/, www.hc-sc.gc.ca/index-eng.php, www.gov.uk/topic/health-protection/infectious-diseases); make it available to the staff in an easy-to-access format. (See Further Reading for suggestions).

Emergency Preparation

Organize a multiagency team with representatives from public health agencies, regulatory agencies, and water utilities and with local political officials to review and exercise existing emergency response plans in the event of a large scale waterborne disease outbreak or other disaster likely to result in waterborne illnesses. Local public health agencies have the primary responsibility for the restoration and protection of health of a community following an outbreak event or other emergency.
   Emergency operational procedures should include the following:

- An emergency notification list that includes phone numbers and e-mail addresses of key persons/agencies that need to be informed about possible outbreaks and that should receive emergency press releases. Every state/province has an emergency management agency and depending upon the scale of the event, it may be useful to coordinate efforts.
- Clear guidelines for household water consumption following an event. For example, boil-water advisories or instructions to drink only bottled water. Statements should be reviewed to ensure current relevance and updated to reflect the most current knowledge.
- A plan for dissemination of information to the public; select a coordination point to which all news media and outside agencies will be directed, and designate one person or one telephone number as the contact. (More than one contact person can create confusion).
- Alternative drinking water sources to be used in cases of emergency and plans for the distribution of this water, if necessary. These include alternative munici-

pal systems, bottled water supplies, portable filtration and/or disinfection devices and home treatment units. (Special attention should be given to backup supplies for hospitals, nursing homes and other places where lack of safe water would be immediately life-threatening).

- Identification of specialty laboratories at the state/provincial and national level that are capable of performing (and willing to perform) procedures not routinely done at local laboratories (e.g., large volume water sampling and processing for pathogenic parasites and viruses, serotyping, electron microscopic examination of stool samples, molecular and immunodiagnostic tests for pathogens in environmental and clinical samples). One or more of these tests may be necessary to identify the causative agents in an outbreak and confirm their transmission route.
- A plan to exercise procedure with tabletop exercises involving all pertinent partners on a regular basis and implement any necessary adjustments based upon review of after-action findings.

Investigate Outbreaks

Notification of a suspected outbreak is often received by health authorities from a laboratory report or a family physician and can be documented on a log such as Form A (Foodborne, Waterborne, Enteric Illness Complaint Report). Public health investigators will then interview cases and persons at risk who are well (controls) to make epidemiologic associations to find a common source. From here a hypothesis can be formed. Further investigation will involve:

- Collecting clinical samples and water samples
- Conducting an on-site investigation at the facility alleged to be responsible to determine the mode of contamination or process failure, e.g., low disinfectant level
- Characterizing the etiologic agents by laboratory analysis using various typing schemes. DNA profiling or pulsed-field gel electrophoresis (PFGE), of isolates from clinical and water samples to "fingerprint" and link strains of the etiologic agent among cases and to the source

Act on Notification of Illness

Prompt handling and referral of water-related complaints, rapid recognition of the problem, and prevention of further illnesses are the foundations of a successful investigation. Complaints of water problems are as likely to be reported to a water utility as to a health department. Communication is essential between these agencies. This first contact with the public is a vital aspect of an investigation of a potential outbreak and needs to begin by public health professionals as quickly as possible, usually within 24 hours, sometimes by putting less urgent activities aside. As indicated earlier, any action respecting a potential deliberate contamination of water will generate a specific approach to further action.

Receive Complaints or Alerts

Upon receiving a complaint or an alert about a water supply or water exposure or illness potentially attributed to these sources, record the information on Form A.

Alerts may be initiated by reports from physicians, laboratories, or from hospital emergency rooms. Alerts may also include an increase in a particular PFGE pattern from clinical isolates. An investigation may be initiated to determine if there is a common water exposure among patients with the PFGE pattern. The pattern may be compared with similar PFGE patterns in PulseNet databases to determine if there are similar occurrences of the pattern in water and clinical isolates nationwide or internationally, e.g., for food that might have been contaminated with water, bottled water. The form provides information upon which to decide whether an incident should be investigated. Form A is not difficult to fill out and can be completed by a public health professional, a trained water utility staff member, or trained office worker.

Assign a sequential number to each complaint. If additional space is needed to record information, use the reverse side or attach additional sheets. Always ask the complainant to provide names of other persons at the event, or using the water supply or recreational water under suspicion, whether or not ill, and names of any other persons who are known to be ill with the same symptoms. Follow up by contacting these additional persons.

Emphasize to the persons making alerts or complaints the need to retain a sample of the suspect water and to save clinical specimens (vomitus and stool) from ill persons using a clean utensil in a clean jar or plastic bag and to seal tightly and label clearly with the name of the person and date, and store in a refrigerator (do not freeze). Also consider family members not ill for case-control studies. Advise complainants to collect a liter (quart) of water immediately, preferably in sterile containers but otherwise in jars that have been boiled or in plastic bags, or if this is not feasible, in other clean containers. Tell the complainant to save any ice cubes or refrigerated water, either in their present containers or in unused plastic bags, in the refrigerator or freezer (if already frozen) where they are normally kept. Instruct the ill person to hold all clinical specimens and water samples until the health agency evaluates the epidemiological evidence and arranges, if necessary, to collect them. If it is determined that the specimen or sample is not necessary, notify the complainant and advise on proper disposal of the material.

Unfortunately, the specific etiologic agent cannot be identified in a large proportion of waterborne outbreaks because water samples and clinical specimens (a) were not collected in an appropriate time-frame (not early enough during illness),

(b) are too old, (c) are too small in volume, especially for *Giardia* and viruses which require liters, (d) have not been examined for the appropriate agent. Contaminants may be in the water system for only a short time, and concentrations of toxic substances and numbers of microorganisms may decrease significantly because of dilution or disinfection.

If there is a cluster of cases, monitor reports from physicians, complaints about water, or records of laboratory isolation of enteric pathogens that may suggest outbreaks of disease or contributory situations. Collect clinical specimens and water as soon as practicable according to the section *Obtain Clinical Specimens* in this book.

Log Alert and Complaint Data

Extract key information (see* and † entries) from Form A and enter it onto Form B (Foodborne, Waterborne, Enteric Illness and Complaint Log). Record time of onset of the first symptom or sign of illness, number of persons who became ill, predominant symptoms and signs, whether ice or water was ingested, and the name of the water supply or recreational water alleged to have caused the illness, and whether a physician had been consulted, and/or had taken feces or emesis samples, and/or prescribed antibiotics. Also, enter on Form B names of the places or common gatherings (other than home) at which the stricken persons ingested water during the 2 weeks before onset of illness (see Table 1 for an example). Enter a code for the water source (e.g., community, non-community, individual well, bottled, stream/ lake, vended, or untreated). Under "history of exposures" column indicate whether the afflicted person had recent domestic or international travel, attended a child care facility, or had recent contact with ill persons or animals. Under "comment" column, enter notations of type of agent isolated, results of specimen tests, places where water was consumed during travel, names and locations of restaurants or other foodservice facilities, and other pertinent information including hospitalization, occupation, or place of employment. At this phase of the investigation it will probably not be known whether the illness is waterborne, foodborne, or person-to-person spread. This log can be kept either in hardcopy or in electronic format. See Table 1 (below) as an example of a log.

Interpretation of Table 1.

Entry 101—Get further details on the patient's symptoms and seek other cases. The report of foreign travel suggests an infection that may have been acquired outside the country. Follow-up of such cases may identify an outbreak of international scope. If so, inform state/provincial and national authorities concerned with surveillance of waterborne disease about the situation.

Entry 102—Possibly food associated; alert food safety officials.

Entry 103—Initiate investigation; the two cases of conjunctivitis suggest the possibility of a common-source outbreak associated with the motel pool.

Entry 104—Initiate investigation; 12 cases indicate an outbreak that has a common time-place association.

Entry 105—This could be related to entry 103, because this person reported swimming in the same pool.

Table 1 Foodborne, waterborne, enteric illness and complaint log

Complaint[a]			Illness			Food		Water			History of exposure[b]	Comments
No.	Date	Type	Onset date	No. ill	Predominant symptoms/signs	Alleged/ suspected	Where eaten within 72 h	Where ingested within 72 h	Where contacted within 2 weeks	Source[c]		
101	8/16	I	8/15	1	Diarrhea			Redguard		C	IT	*Giardia* isolated from stool
102	8/23	I/UE	8/22	1	Diarrhea	Roast beef	Speedy Foods: Joe's Diner	Dixon		C		
103	8/23	I/RW	8/22	2	Conjunctivitis			Plainville	Shadygrove Motel pool yesterday	C		
104	8/23	I	8/22	12	Diarrhea		Dixon day care center	Dixon			CC	
105	8/25	I	8/25	1	Conjunctivitis			Plainville	Shadygrove Motel	C		
106	8/26	I	8/26	1	Diarrhea, fever			Hampton		W		*Escherichia coli* O157 isolated from stool
107	8/26	RW		0				Lake Orly			NC	Water smelled like rotten eggs
108	8/27	I	8/27	1	Facial flushing, dizziness	Tuna fish	Fred's Café	Dixon		C		
109	8/29	I	8/27	2	Bloody diarrhea					S	DT	Drank stream water while hiking
110	8/29	DW	0					Private well		W		Rust and silt in water

(continued)

Table 1 (continued)

No.	Date	Type	Onset date	No. ill	Predominant symptoms/signs	Alleged/ suspected	Where eaten within 72 h	Where ingested within 72 h	Where contacted within 2 weeks	Source[c]	History of exposure[b]	Comments
111	8/29	I	8/29	1	Diarrhea, abdominal cramping			Hampton		W	A	*Escherichia coli* O157 isolated from stool
112	8/31	I	8/30	1	Baby bluish color			Farm well		W		Ill 2 weeks after moving to farm
113	8/31			0				Redguard		C		Total coliform count 100/100 mL
114	8/31	I	8/28	6	Itching red papules on skin			Dixon		W/L		Swam in Brown's farm pond
115	9/7	CF		0		Baby cereal		Dixon		C		Metal fragments found in Brand X, Lot JK20-111E
116	9/8	DW		0						C		Heavy rains, flooding, turbid water at Winding Bend area of town

Header spans: Complaint[a] (No., Date, Type); Illnesss (Onset date, No. ill, Predominant symptoms/signs); Food (Alleged/suspected, Where eaten within 72 h); Water (Where ingested within 72 h, Where contacted within 2 weeks, Source[c]); History of exposure[b]; Comments

[a]Type of complaint: I illness; CF contaminated/adulterated food; UE unsanitary food establishment; DW poor quality drinking water; RW poor quality recreational water; MP complaint related to media publicity; D disasters; O others

[b]Exposure history: DT domestic travel (out of town, within country); IT international travel; CC child care; CI contact with ill person outside household or visitor to household; A An Exposure to ill animal; C contact with ill person within household

[c]Water source: C community; NC non-community; W well; B bottled; S/L stream/lake

Entry 106—This entry and that of entry 111 could have a common source; investigate. quality of water.

Entry 107—This entry and those of 110 and 116 indicate the possibility of substandard water. Either advise callers or refer their complaints to someone who can (e.g., the water utility), but stay alert and check for illness in communities where these situations occurred.

Entry 108—Possibly food associated; alert food safety officials.

Entry 109—Initiate an investigation; the situation suggests a common source outbreak.

Entry 110—See entry 107.

Entry 111—See entry 106. Could also be exposure to an animal.

Entry 112—The syndrome suggests methemoglobinemia. Sample water and test for nitrites/nitrates.

Entry 113—Resample and investigate to find the likely cause of the elevated coliform count.

Entry 114—The syndrome suggests an outbreak of water-contact infection, possibly swimmer's itch.

Entry 115—Possibly food associated; alert food safety officials.

Entry 116—See entry 107.

Review the log each time an entry is made and also each week to identify clusters of cases and/or involvement of a common exposure that might otherwise go undetected. If your agency has district offices or if there are nearby jurisdictions (as in metropolitan areas), periodically send copies of log sheets to a central coordinating office (e.g., weekly or when there are 10–20 entries). Reports of current illness levels should include historical information on illness trends in the community so that new data can be considered in the appropriate context. Report to your supervisor if you suspect any time, place, or person associations and take steps to initiate an investigation.

Refer Complaint to Proper Agency

Refer complaints that fall outside your agency's range of operations to the appropriate authority, such as the Department of Health, Ministry of the Environment, and indicate the action taken in the disposition box on Form A. Develop a working relationship with such authorities so they will reciprocate in situations which may be associated with illness. Often an investigation requires efforts of more than one agency. Cooperation and prompt exchange of information between agencies are vital.

Prepare for the Investigation

Prior to conducting investigations, personnel should know the surveillance protocol, and be trained on how to develop questionnaires, conduct interviews, and use investigation related software. All trained investigative team members should be assigned a role and the person heading the investigative team, should "be made" responsible for the investigation, if this was not done when the surveillance proto-col was established. Delegate sufficient authority and provide resources to the head investigator so that the investigation tasks can be accomplished effectively and efficiently. Inform all outbreak investigative team members that any findings are to be reported to this delegated authority. A list of all team members and additional contacts such as administrative contacts, sanitarians/environmental health officers/ public health inspectors, local and regional contacts, physicians, clinical laborato-ries, or other persons who may become involved in outbreak investigations should be assembled.

Before beginning the investigation, check the supply of forms and the availabil-ity of equipment suggested in Table A (Equipment useful for investigations) and obtain any needed materials or additional equipment. General resource materials describing signs and symptoms, incubation times, and specifics regarding speci-men collection and appropriate kits to be used should be maintained and readily available to those processing the initial calls, which may help to formulate the initial hypothesis.

If the alert or complaint suggests a possible outbreak, inform laboratory personnel of the type of outbreak and estimated quantity and arrival time of clinical specimens and water samples collected. This information will give laboratory managers time to prepare laboratory culture media, prepare reagents, and allocate personnel. At a minimum, the laboratory should have six to eight stool culture kits on hand or read-ily available, since in many cases, stool specimens must be collected within 72 h of onset of illness to isolate and identify certain pathogens (e.g., *Campylobacter* spp., and *Salmonella* spp.). Consult laboratory personnel about proper methods for col-lecting, preserving, and shipping environmental samples and clinical specimens if such information is needed. Obtain appropriate specimen containers and sample submission (chain of custody forms) from them.

Once the investigation is underway, the proper clinical specimens should be col-lected as soon as possible before patients recover and become less likely to submit specimens. All suspected waterborne outbreaks should be examined and a determi-nation made regarding the feasibility of conducting a thorough investigation even if the time to collect proper clinical specimens has passed.

Verify Diagnosis

An ill person or family member, physician, hospital staff member, or operator of a water utility or recreational site may report suspected cases of waterborne illness. Whatever the source of the report, verify the diagnosis by taking a thorough case his-tory and, if possible, by reviewing clinical information and laboratory findings. (This analysis can be further substantiated by detecting suspected etiologic agents in water). Verification of the diagnosis is done in consultation with medical professionals.

Get Case Histories

When a complaint involves illness, complete Forms C1–2 (Case History: Clinical Data and Case History: Food/Water History and Common Sources) either at the time of initial notification, during a personal visit, or during a telephone call to the person reported to be ill. Use this same detailed interview approach with every person who has been identified in the initial complaint or alert, even though some may not have been ill. Be aware that potential cultural and language barriers can make interviews difficult. A different interviewer may be needed to accommodate these barriers. Continue this until sufficient information is obtained to decide whether there is, indeed, an outbreak of waterborne illness. From persons who are at risk of illness but who remained well, also obtain water and 72-hour food histories, inquire about recreational water exposure in past 2 weeks, and information about their activities in common with the ill persons. Information from these persons is as important to make epidemiologic associations as it is from the cases.

When it is apparent that an outbreak has occurred and a specific event has come under suspicion, substitute Forms D1–2 (Case History Summaries: Clinical Data and Case History Summaries: Water/Laboratory Data) for Form C. Form D1 can be used initially in many routine waterborne illness outbreak investigations where it is obvious that a common-source outbreak has occurred or when all of the ill persons consumed water together (e.g., drank from the same public system, consumed ice at an event) or recreated at the same place (e.g., swam in the same lake or used the same hot tubs). This will simplify recording, because most affected persons will give similar information. At this time, notify the district, state, or provincial epidemiologist about the outbreak.

If a specific pathogen (e.g., norovirus, *E. coli* O157:H7, *Cryptosporidium* spp.) has been identified as the etiologic agent, consider developing a form for recording relevant information. Many state/provincial or national public health agencies have standard forms tailored to specific pathogens. Include signs and symptoms of the illness and other clinical information, the etiology of the agent, and usual methods of transmission. Computer programs (e.g., Epi Info™) can aid in the design of such standard forms.

Upon contact with the affected person, identify yourself and your agency and explain the purpose of the visit or call. A professional attitude, appropriate attire, friendly manner, and confidence in discussing epidemiology and control of waterborne illnesses are essential for developing rapport with affected persons or their families and in projecting a good image of the investigating agency. Keep in mind that you are not interviewing someone you inspect or regulate, but that you are providing a service to the affected person. Exhibit genuine concern for persons affected and be sincere when requesting personal and confidential information.

Communicate a sense of the urgency of the investigation, and emphasize that their participation will make a positive contribution for the control and prevention of waterborne illness. Parental consent must be obtained before interviewing children under 18 years of age. In some locations, consent from the affected person's physician may also be required.

After asking open-ended questions about the person's food exposures and illness history, follow up with more specific questions to fill in the details and better ensure a thorough recall. Base your level of communication on a general impression of the person being interviewed, considering information about age, occupation, education, or socioeconomic status. Tact is essential. Use either Form C or Form D, as appropriate, as a guide. State questions so that the persons who are being interviewed will describe their illnesses and associated events in their own words. Try not to suggest answers by the way you phrase questions.

Fill in Form C1–2 (if appropriate) and take additional notes during the interview. Ask specific questions to clarify the patient's comments. Think questions through before conducting the interview. Realize that people are sometimes sensitive to questions about age, sex, special dietary habits, ethnic group, excreta disposal, and housing conditions. Nevertheless, any or all information of this type can be relevant. Word questions thoughtfully when discussing these characteristics and habits. Such information can often be deduced from observations. If doubt remains, confirm your guesses by asking indirect questions. Information on recent travel, gatherings, or visitors may provide a clue to common sources or events that would otherwise be difficult to pinpoint. Review known allergies, recent immunizations, recent changes in the patient's medical status, and similar information. Remember that the agents associated with waterborne disease can also be spread by other means such as consuming food, person-to-person, visiting child care centers, animal-to-person in petting zoos, through walk-in-spray fountains, and pools for young children.

As persons describe their illnesses, check boxes next to appropriate symptoms or signs on Form C1. Do not ask about all symptoms or signs listed; however, ask about those marked with an asterisk if the ill person does not mention them. If there are questions, explain symptoms to the patient in understandable terms. The symptoms and signs in the first two columns of Form C1 are usually associated with poisoning or intoxication, although some occur during infections. Those in the third, fourth, and fifth columns are usually associated with enteric infections, generalized infections, and localized infections, respectively. Those in the last column are usually associated with disturbance of the central nervous system.

Diseases in any category will sometimes be characterized by a few symptoms and signs listed in the other columns, and not all signs and symptoms occur for any one ailment or for all persons reporting illness. If an illness seems to fall into one of these categories, mention other symptoms in the category and record the patient's response.

Whenever possible, use physician and hospital records to verify signs and symptoms reported by patients. Clinical data may strengthen or dismiss the possibility of waterborne illness. Before contacting a physician or a hospital, become familiar with laws and codes relating to medical records to ensure that you have legal access to these records. Legal release forms may be necessary to obtain some records. Do not distribute names of patients, their other personal identities (e.g., address, phone number), or their clinical information to unauthorized persons.

The entries begin with the day of illness, followed by the previous 2 days. If the illness, however, began early in the day or before any of the listed meals, modify the entries on the form so that the 72-hour history can be completed in the space provided on the form. If the incubation period is 3 days to a week in duration, use additional copies of Form C2 and modify day or day before subtitles.

Signs and symptoms will sometimes give clues to the transmission route by indicating the organ systems affected. If the early and predominant symptoms are nausea and vomiting, ask about the most recently ingested water or beverage within the past 6 h. In these situations, suspect high-acid water supplies, carbonated beverages and fruit drinks, because these tend to leach metallic ions from water pipes and containers. If diarrhea, chills, and fever predominate, be suspicious of water and beverages ingested 12–72 hours before onset of illness for salmonellosis, shigellosis, and norovirus related gastroenteritis. If the incubation period averages 1–2 weeks, consider typhoid fever, cryptosporidiosis or giardiasis. Diseases with incubation periods exceeding 2 weeks (e.g., hepatitis A and E, amebic dysentery, or schistosomiasis) can be handled as special cases for which longer histories would be sought. Others, such as chronic lead and arsenic poisoning, have incubation periods of variable durations and onsets so gradual as to be indeterminable. See Table B (Illness acquired by ingestion of contaminated water: A condensed classification by symptoms, incubation periods, and types of agents) for details on specific pathogens, Table C (Illnesses acquired by contact with water: A condensed classification by, symptoms, incubation period, and types of agents), and Table D (Illnesses acquired by inhalation of microorganisms aerosolized from water. A classification by symptoms, incubation period, and type of agent).

Other microorganisms not listed in Tables B, C, and D that can be potentially spread by water include the bacteria *Klebsiella pneumoniae*, *Mycobacterium avium* complex, *Acinetobacter calcoaceticus*, *Elizabethkingia meningoseptica*, *Stenotrophomonas maltophilia*, *Pseudomonas putida*, *Serratia marcescens*, protozoa *Isospora*, *Microsporidium*, algae *Schizothrix calcicola*. These microorganisms are less frequently identified with waterborne illness, but they may become opportunistic pathogens, particularly for highly susceptible and immunosuppressed persons. Further investigation is needed to confirm their role in the spread of waterborne diseases.

Gather information about all sources of water to which the patient(s) may have been exposed 2 weeks before onset of illness. The water supply and the event that precipitated the illness might not be obvious. Persons often have difficulty recalling exposure to all water sources including; ice or water ingested; aerosols and recreational water contact. Therefore, if the person does not remember specific exposures to water, ask about the water consumed in usual or routine daily habits and the amounts ingested; exposure to recreational waters; and unusual exposures or events attended during this interval. This may stimulate recall of away-from-home water consumption or contact that was unusual. Ask about other risk factors for enteric illness, such as contact with young children and child care centers, animal contact, ingestion of raw foods of animal origin, and usual food preference habits.

For persons who have been traveling, ask them where (both cities and rural areas) they have traveled during the incubation period of suspected agents. Determine if they drank water from any taps or pumps in rural areas they visited. Ask whether unheated (or untreated) tap water or beverages containing unheated (or untreated) water or ice was ingested at restaurants, in hotels or at events in the places they visited. Also, ask whether they ingested bottled water and, if so, the brand name. Find out whether they drank water from streams, ponds, springs, or other natural water sources. If they did, ask if they observed any abnormal condition of the water such as algal growth, high turbidity or discoloration. Ask if domestic or wild animals had access to the water.

If they have skin or eye infections or generalized infections, ask them to name all swimming pools, water slides, beaches, lakes, ponds, or other chlorinated and non-chlorinated water courses where they swam, waded or bathed during their trip. Also ask them whether they used any hot tubs, spas, whirlpools, or similar devices. This information sometimes provides clues to common sources or to events that otherwise would be difficult to discover. Record the information on Form C1.

In a protracted outbreak, or when investigating an outbreak of a disease with a long incubation period, expect recall to be poor. In this situation, obtain from ill persons and others at risk a listing of their water, ice, and beverage preferences and amounts usually ingested, or their purchases of these items within the range of the incubation period of the suspected disease. As a guide, draw up a list of either water, ice, and beverages that are commonly consumed by the affected group or those waters, ice, and beverages previously identified as vehicles of the suspected disease under investigation. Summarize data from all copies of Forms C1–2 on Form D. Form D allows rapid review of all exposed persons (ill or not ill) and serves as a basis for analyzing the data.

Obtain Clinical Specimens

Diagnosis of most diseases can be confirmed only if etiologic agents are isolated and identified from specimens obtained from ill persons. Get specimens from the ill persons to confirm an etiologic agent.

- In large outbreaks, obtain fecal specimens from at least ten persons who manifest illness typical of the outbreak
- In smaller outbreaks, obtain specimens from as many of those ill and those at risk as practicable, but from at least two, and preferably ten, ill persons
- Try to collect specimens before the patient takes any medication. If medication has already been taken, collect specimens anyway, and find out the kinds and amounts of medicine taken and the time that each dose was taken
- Also get control specimens from persons with similar exposure histories that did not become ill

Obtain clinical specimens at the time of the initial interview during acute illness or as soon as practicable thereafter. Even though this is not always possible, take specimens even after recovery because etiologic agents may remain in low populations or concentrations. If a disease has already been diagnosed, collect specimens as listed in Table B. If a disease has not yet been diagnosed, choose kinds of specimens that are appropriate to the clinical features. Laboratory information obtained from the first patients may be useful to physicians in the treatment of cases detected later. Apart from the fact that people are more likely to cooperate while they are ill, some pathogens or poisonous substances remain in the intestinal tract for only a day or so after onset of illness. If the patient is reluctant to provide a fecal specimen explain that the specimen will be tested to identify the causative agent and compare it to any agent recovered from the water.

If a disease has not yet been diagnosed, choose specimens that are appropriate to the clinical features. Laboratory information obtained from the first patients may be useful to physicians in treating cases detected later. Some pathogens (e.g., *Salmonella*, parasites) may be recovered for weeks after symptoms have abated. If applicable for the disease under investigation, take specimens even after recovery because some etiologic agents may remain in low numbers, and changes in serologic titers can be detected.

Before collecting specimens, review Table E (Guidelines for specimen collection) and, if necessary, get additional instructions from laboratory personnel and seek their advice on how to preserve the stool specimens if you cannot deliver them to the laboratory immediately. Many public health agencies have special fecal specimen kits. Demonstrate to the patient how to use the materials in the kit, how to complete the form in the kit and how to mail it if you are not going to pick it up. If mailing specimens, make sure that you are aware of the regulatory requirements that may apply to the transport of infectious material.

Stool specimen containers for intestinal parasite examination are not suitable for bacterial or viral examinations because they ordinarily contain a preservative, such as formalin or polyvinyl alcohol. If an inappropriate transport medium is used, a specimen can be rendered unsuitable for laboratory examination.

Feces. If the patient has diarrhea or is suspected of having had an enteric disease, obtain a stool specimen (preferred specimen) or a rectal swab. Instruct patients to provide you with their own specimens by one of the following means.

1. If practicable, give the patient a stool specimen container with a wooden or plastic spoon or a tongue depressor. A clean container available in the home (e.g., a jar, or disposable container that can be sealed) and a clean plastic spoon or similar utensil can be used if laboratory containers are not available.
2. Label the specimen container with the patient's name age/date of birth and date of collection.
3. Collect the stool specimen by one of the following methods:
    (a) Put sheets of plastic wrap or aluminum foil under the toilet seat and push them down slightly in the center, but not so far as to touch the water in the bowl. Sheets of paper can be tacked on the rise of a latrine and pushed down to form a depression in which to catch feces. Take care to ensure that toilet cleaning chemicals and other microorganisms in the toilet bowl do not contaminate the fecal specimen. After defecating, use a clean spoon or other utensil to transfer about 10 g of feces into a specimen container or other clean container.

(b) Defecate directly into a large clean dry container or bedpan. Use a clean spoon or other utensil to transfer about 10 g or the size of a walnut of feces into a specimen container or other clean container.

(c) Scrape feces off a diaper with a clean spoon or other utensil to transfer about 10 g of feces into a specimen container or other clean container.

4. Collect fecal swabs by twisting the cotton-wrapped end of the swab into the stool obtained in one of the ways described above. Follow instructions given in Table E. If necessary, use fecal-soiled toilet paper or cloth diaper and twist a swab into the top of feces. Take care to ensure that there is no carryover of toilet paper as they are impregnated with barium salts which are inhibitory to some fecal pathogens.

Dispose of excess fecal material into the toilet and carefully wrap all soiled articles (e.g., by placing them inside two plastic bags) and dispose of in domestic waste. Check that the specimen container is tightly sealed and properly labeled and place into a clean outer plastic bag (special zip lock bags for clinical specimens, if available). Store the specimen in a cool place, preferably at 4°C to await pick-up or despatch. DO NOT FREEZE.

Feces from Rectal Swabs. Collect rectal swabs by carefully inserting the swab approximately 2.5 cm (1 in) beyond the anal sphincter. Gently rotate the swab. Fecal matter should be evident on the swab.

Vomitus. If the person is vomiting or subsequently does so, arrange to collect vomitus. Tell the patient to vomit directly into a sterile specimen container or a plastic bag. Otherwise, transfer some vomitus from a clean receptacle into the container with a clean spoon. Refrigerate, but DO NOT FREEZE, this specimen until it can be picked up or delivered to the laboratory.

Blood. Take blood if a patient has a febrile infection or when infectious agents are suspected (see Tables B, C, and D). Blood specimens are collected for:

- Bacterial culture
- Detection of antibodies to specific agents
- Detection of certain toxins

Before collecting specimens, get additional instructions from laboratory personnel and seek their advice. Blood should be obtained by an appropriately trained and accredited person (check appropriate laws). Collect blood during the acute phase of illness, as soon as the febrile patient is seen (within a week after onset of illness) and, if comparing of serologic titers, again within 6 weeks (usually 2–4 weeks later) during the convalescent phase. Draw 15 mL of blood (from an adult) or 3 mL (from a child) or 1–2 mL (from an infant). If possible, collect the blood from the same patients from which stool specimens were obtained if both specimens are to be examined. Label tubes and vials at every step of serum transfer. DO NOT FREEZE whole blood because the resultant hemolysis interferes with serologic reactions.

Blood for culture (for pathogens such as invasive *Salmonella* species, *Vibrio vulnificus*)	Inoculate freshly collected blood into culture bottle supplied by the laboratory
Blood for detection of – Antibodies (to pathogens such as *Salmonella* Typhi, hepatitis A virus, *Toxoplasma gondii*) – Toxins	Collect into a sterile syringe or evacuated sterile tube that does not contain anticoagulants. If practicable, centrifuge the blood at 1,000 rpm for 10 min; pour off the serum into small screw-cap vials and store at approximately −18°C. If the serum cannot be separated immediately, rim the clot with a sterile applicator stick and refrigerate approximately 4°C to get maximum clot retraction if the specimen is to be stored unfrozen overnight. If centrifugation cannot be done, store the blood specimens in a refrigerator until a clot has formed, then remove the serum and transfer it with a Pasteur pipette into an empty sterile tube. Send only the serum for analysis

Urine. Instruct patients to collect urine in the following manner. Clean the area immediately around the urethral orifice with a paper pad that has been pre-moistened with 4% tincture of iodine or other appropriate antiseptic. Then begin to urinate into a toilet and collect 30 mL (about 1 oz) of midstream urine into a sterile bottle. Use either a second antiseptic-moistened pad or an alcohol-moistened cotton ball or tissue to clean any drops from the top or side of the bottle.

Other Instructions. Follow applicable instructions given in Table E. Before or immediately after collecting clinical specimens, use waterproof permanent markers to label each container with the patients name, complaint number, case identification number, specimen number, date and time of collection, tests requested, and other appropriate information. Tightly seal all containers.

Clinical Specimen Collection Report for each specimen. Complete Form E (Clinical specimen collection report). The complaint number, case identification (ID) number, and specimen number must be entered on each report so that laboratory results can later be correlated with other data. On Form C1 record the type of specimen collected, and submit both the specimen and a copy of Form E to the laboratory. Send a copy of the laboratory report to the patient's physician or call if urgent.

Pick Up Water/Ice Samples and Containers that the Patient Collected

If the patient/case or other household member collected any water, ice, or beverage as instructed during initial contact, label containers with the complaint/outbreak and sample numbers. Proceed as instructed in Table F (General Instructions for collecting water samples for microbiological analysis) and complete Form F (Water/Ice collection report) and/or Forms G3–G8 as applicable. Record conditions of collection as called for on the forms. If a hypothesis associates the illness with water, caution these persons not to use the water source unless the water is first boiled and to discard all previously prepared ice and water-containing beverages until notified otherwise.

Develop a Working Case Definition

Develop a working case definition to classify exposed persons as either cases or non-cases. Start with the most specific symptoms (such as diarrhea and vomiting) rather than broader symptoms such as nausea or malaise. For example in an outbreak of gastroenteritis, a case might be defined as a person from whose stool a specific pathogen was isolated. It may be a person who was at risk and developed diarrhea within a specified period of time. Diarrhea will have to be defined, perhaps as three or more loose, watery stools during a 24-hour period. In some cases, a particular pathogen responsible for the outbreak might have been identified from clinical specimens. A case definition, which is developed later in the investigation, might include either a person having specific signs and/or symptoms within a period of time or a person from whom a specific pathogen was isolated. The ultimate case definition has a tremendous impact on the investigator's ability to make illness and exposure associations and to calculate probability of these associations.

Sometimes the first symptom or sign provides a clue to developing a case definition. Information in Tables B, C, D, G, and H can be useful in making case definitions. Compare newly identified cases with the definition to see whether each is part of the outbreak.

Classify cases into categories:

- A **confirmed case** is a person with signs and symptoms that are clinically compatible with the disease under consideration and for which there is either (a) isolation of an etiologic agent from (or otherwise identified in) an appropriate specimen from the patient, or (b) serologic evidence of a fourfold or greater rise in convalescent antibody titer. A confirmed case must also have possible exposure to the etiologic agent within the incubation period of disease. See Table E.

  Criteria for confirmation of etiologic agent responsible for outbreaks of waterborne illnesses for definitions of confirmed cases for specific waterborne diseases:

- A **presumptive case** is a person with signs and symptoms that are clinically compatible with the disease under consideration, and for which there is laboratory evidence of infection (e.g., an elevated antibody titer but less than a fourfold increase), but the etiologic agent was not found in specimens from patients or no specimens were collected. A presumptive case must also have possible exposure to the etiologic agent within the incubation period of disease.

- A **suspected case** is a person with signs and symptoms that are clinically compatible with the disease under consideration and history of possible exposure, but laboratory evidence is absent, inconclusive or incomplete.

- A **secondary case** is a person who became infected from contact with a primary (outbreak-associated) case or from a vehicle contaminated by a primary case. Onset of illness for secondary cases typically is one or more incubation periods after the outbreak-associated cases.

It is not essential, however, to classify cases into these categories. Do so only if it aids in developing a final case definition or in making comparative analyses of data. Consider doing analyses using case definitions of both confirmed and combined confirmed, presumptive, and highly suspect cases, and compare the results.

Make Epidemiologic Associations

Make a preliminary evaluation of the data collected as soon as possible. If you decide that there is an outbreak, use the information you have to develop a hypothesis about the causal factors.

Determine Whether an Outbreak Has Occurred

An outbreak is an incident in which two or more persons have the same disease, have similar clinical features, or have the same pathogen (thus meeting the case definition), and there is a time, place, or person association among these persons. A waterborne outbreak is traceable to ingestion of contaminated water or ice or to contact with contaminated water.

A single case of either chemical poisoning or a disease that can be definitely related to ingestion of drinking water or contact with water can be considered an incident of waterborne illness and warrants further investigation. Waterborne methemoglobinemia in an infant who resides in a rural area having a high concentration of nitrates in well water is an example of a single case of waterborne illness due to ingestion. A rare diagnosis such as primary amebic meningoencephalitis in a person who swam in a body of freshwater and inadvertently ingested the ameba, *Naegleria fowleri*, through the nose is an example of a single incident related to water contact.

Sometimes it will be obvious from an initial report that an outbreak of waterborne disease has occurred simply because of the number of persons displaying certain signs and symptoms at or near the same time. Many complaints, however, involve illness in only one or a few persons. It is often difficult to decide whether ingestion or contact with a particular water source and onset of illness was associated or coincidental. Certain diseases that are highly communicable (e.g., shigellosis and epidemic viral gastroenteritis) may result in secondary infections from person-to-person spread or from subsequently contaminated food or water.

However, if complaints are received from several persons who are associated with ingesting water or contact with water at the same place, water is likely to be involved. Routine review of the log pertaining to potential waterborne illnesses for similar complaints can often be useful in detecting time, place or person associations. An investigation may also proceed based upon the suspicion of an intentional contamination of a water source.

Make Time, Place, and/or Person Associations

A time association exists if the time of onset of similar illnesses is within a few hours or days of each other. Place associations exist when persons have ingested water from a particular single source, have swum in, worked in or otherwise been exposed to the same water, have attended the same event, or reside in an area common to all. Person associations indicate a shared personal characteristic, such as being of the same age group, sex, ethnic group, occupation, social group, or religion. Waterborne illnesses transmitted by a community water supply usually afflict persons of both sexes and all ages throughout the community. Non-community water sources, such as bottled water, ice, water from individual wells, or water from areas of recreation should also be considered when making associations. Keep in mind that water can contaminate foods during washing or freshening, and it can contaminate utensils and vessels that are used to handle or store foods. Water may therefore be a source of contamination of another vehicle. Also, water can be ingested as aerosols generated by shower heads, whirlpools, hot tubs, fountains, cooling towers, and irrigation devices. Once some of these associations become obvious, question other persons who could be at risk because of their time, place, or person associations with the ill persons.

Formulate Hypotheses

From time, place, or person associations that have been established or suggested by the investigation, formulate hypotheses to explain (a) the most likely type of illness, (b) the most likely vehicle involved, (c) where and the manner by which the vehicle became contaminated, and (d) other possible causal relationships. The section "Collection and Analysis of Data" describes calculations that can aid in the formation of these hypotheses. Test hypotheses by obtaining additional information to support or reject them. If the hypothesis includes food contamination, the instructions given in the manual, *Procedures to Investigate Foodborne Illness*, might be useful. Guidelines for confirmation of waterborne outbreaks are presented in Table G and Guidelines for confirmation that water is responsible for illness are presented in Table H.

Possible Precautionary Control Actions

If there is strong evidence to support a hypothesis that the outbreak is waterborne, take precautionary actions. The choice of action is dictated by the (a) suspected causal agent, (b) size of the water source, (c) availability of alternate water sources, and (d) expected use of the water. On the basis of available information, estimate the population at risk and engage any public relations staff with your organization to help inform all persons potentially impacted.

When dealing with a microbiological contaminant or agent, consider issuing a boil-water advisory with water treatment guidelines (e.g., heating water in a covered container to a rolling boil for at least 1 min and keeping it covered until use). Other options that can be explored include chlorinators that can be installed in individual and non-community systems. For community and non-community supplies in which chlorine is already used, increasing the chlorine dosage and opening hydrants and taps to draw the super-chlorinated water through the whole system might be an option. Increasing chlorine is sometimes not effective because the chlorine contact time is too short or super-chlorinated water does not reach some parts of the system. Furthermore, chlorination is ineffective against *Cryptosporidium* oocysts and requires a long contact time to kill other human pathogens like hepatitis A virus and *Giardia*.

For suspected chemical contamination contact a specialist for further assessment and remedial strategies, such as activated charcoal filters. As a last resort, shut off the contaminated system until the source of contamination is found and controlled. Be cautious when you take this drastic measure, because it may do greater harm than good by causing lack of water for hospitals, nursing homes, or for firefighters to extinguish fires. If the water is shut off or the treatment facility or distribution system disrupted (as in the case of floods or other disasters), consider means to distribute water from an alternative source to healthcare facilities and homes.

If an illness could have resulted from water contact, close the offending water source, post warning signs around it, and patrol the area. Where there is a swimming pool, hot tub, spa, fountain, or whirlpool, evaluate the recirculation system and its operation. It may be that increasing disinfectant concentration by super-disinfection could resolve the problem. Where there may be chronic operational problems, evaluate pH, disinfectant concentration, and bacteriological laboratory records. Choose your course of action, including consultation with appropriate professional experts, depending on the contributing factors existing at the time of investigation.

Verify the effectiveness of these actions (e.g., boil-water advisory, super-chlorination, provision of alternate water source) to protect public health by monitoring illness levels in the population to determine if the outbreak terminates. If the outbreak continues unabated, consider the possibility of other transmission routes. Also, verify the effectiveness of repairs to the water system, super-disinfection, and other actions by closely monitoring the quality of the water supply or recreational water to determine if laboratory reports indicate that the water is now safe for consumption or contact.

Inform the Public

If there is a public health threat, work with any available public relations staff to announce the outbreak in the mass media so that the public who consumed or was otherwise exposed to the implicated water can be alerted to take appropriate action including seeking medical consultation or treatment. Provide only objective factual information about the outbreak. Coordinate among the investigating agencies to assure that a consistent and accurate message is delivered. It is easy for agencies to miscommunicate before and during a water crisis (See Box 1, False Alarm; Box 2, The Walkerton Outbreak; Box 3, The Flint Water Crisis). It is often preferable to have one spokesperson for all agencies. **Do not release preliminary information that has not been confirmed.** The person giving information about an outbreak should be well informed about the etiologic agent being investigated and prepared to deal with questions. If the health hazard warrants a public warning at the hypothesis stage, tell the public why emergency measures are being invoked and that subsequent information may be cause to modify the action. As the investigation proceeds and the etiologic agent is confirmed and contributory factors are identified, consider terminating emergency measures, and give advice on specific control and preventive measures. Attempt to reach all segments of the population at risk; this may require communication in multiple languages. Route all news releases or statements to all persons involved in the investigation. In situations involving large outbreaks or highly virulent or toxigenic etiologic agents, set up an emergency hotline for the public to call to ask questions. This is likely to occur if there is an intentional contamination event where there is high publicity and public concern. Train staff to handle these calls in a consistent manner so that the advice is the same who gives it. Faulty information derived from poorly tested hypotheses can lead to severe political, legal or economic consequences. An example of this occurred in Sydney, Australia, in 1998 when an apparent water contamination event was publicized for the public to take precautionary actions. The false alarm was costly because of rebates to water customers, additional water testing, and for hiring extra staff, as well as a loss of confidence in the facility (see Box 1, False Alarm). They may then be disseminated by the mass media with inappropriate interpretations of the public health significance. Furthermore, this information may be used as an unrealistic base for water programs or water regulations because of either misinterpretations or pressure from misinformed consumer–advocate groups. All involved parties should follow a written protocol for cross-agency communication and release of information to the public. Unreasonable delays are unacceptable.

Expand the Investigation

Test hypotheses by obtaining additional information to either confirm or refute their validity. Do this by case–control or cohort studies, additional laboratory investigations, and on-site investigations (e.g., laboratory reports of water testing).

Box 1 False Alarm: The Impact of a Poor Water Pathogen Oversight System

Sydney Water (a New South Wales state-owned corporation) supplies 1600 million liters of water each day to 1.5 million properties in Sydney and its outlying areas. The city has a large and complex catchment with nine major dams and several storage reservoirs. About 21,000 km of water main, almost 200 pumping stations and many tunnels deliver water from four main river systems. The water is filtered through eleven treatment plants. Seven are owned by Sydney Water and four are privately owned. These plants provide 90% of Sydney's drinking water and one plant, Prospect, provides up to 80%. In 1998, the quality of Sydney's drinking water came under acute review when *Giardia* and *Cryptosporidium* were found in the city's main water supply at the Warragamba Dam. Initially, low levels of these parasites were first detected in the water supply on 21 July, but these were within the acceptable health limits. In days following, much higher levels were recorded, and on July 27 the first "boil water" alert (in which residents were instructed to boil their tap water before use) was declared for the eastern Central Business District of Sydney. However, by late on July 29 high readings were found in samples at the Prospect Filtration Plant, in a reservoir and at a location further down the system, and a "boil water" alert was issued for the south of Sydney Harbour, and on July 30 a Sydney-wide "boil water" alert was issued affecting most of Sydney's residents. On August 4 the warning was discontinued. However, high levels were again found on August 13 (the second event), although it was believed that most organisms would likely be dead. More positive readings were found on August 14, although at lower levels. Further contamination was identified on August 24 and an extended boil water alert was again declared. This was progressively lifted suburb by suburb until further contamination was reported on September 5 (the third event). A 2-week alert was then instituted, which was finally lifted on September 19. It was determined that the parasitic contamination was caused by low-quality surface water entering the dam. This contaminated source was attributed to moderate rainfall in July, followed by heavy rainfall in August and September which caused intermittent supplies of the raw water to enter the dam. Despite high levels of *Cryptosporidium* (up to >12,000 oocysts) and *Giardia* (up to >7600 cysts) being recorded in July and August, 1998, no increase in human cryptosporidiosis or giardiasis was detected in the exposed population.

The incident was highly publicized and caused major a public alarm because the number of people affected, the on and off boil water alerts, and the fact that the filtration plant had been advertised as one of the best in the world. The economic and political repercussions were extensive. The cost of the crisis to Sydney Water was estimated at A$33 million which included $20 million paid in rebates to customers, $13 million in lost revenue, water testing and staff costs and at least $2.5 million for damages claims. These costs exclude those relating to improvements to the system and infrastructure. The lack of cases of cryptosporidiosis, giardiasis or other water-related health problems led to suggestions that the parasites were either not an infectious type, or not as extensively distributed. An inquiry after the event revealed the publicity as an exaggeration of fact, with Australian Water Technologies, part of Sydney Water, severely overestimating levels of *Cryptosporidium* and *Giardia* present in the water, with the recorded levels exposed to consumers as not harmful to human health. The handling of the crisis by State-owned Sydney Water was heavily criticized, causing the resignation of both the chairman and the managing director, and bringing up issues of private vs. public ownership and scientific uncertainty. The eventual consequence of the State Inquiry was the establishment of the Sydney Catchment Authority in 1999 to assume control of Sydney's catchments and dams, while Sydney Water maintained responsibility for water treatment and distribution and for sewage collection, treatment and disposal.

Obtain Assistance

If an outbreak investigation requires resources beyond your agency's capacity, request assistance from other health professionals. It is desirable to have a team including, if feasible, an epidemiologist, an engineer, a microbiologist, a sanitarian/environmental health office/public health inspector, a chemist, a physician and others, to undertake a detailed waterborne illness investigation. Such personnel can usually be provided by local, state/provincial, or national agencies concerned with health, environment, or agriculture, depending on the expertise needed. For events suspected to arise from intentionally contaminated food, contact emergency response or law enforcement agencies.

Find and Interview Additional Cases

Continue to search for and interview both ill persons who have had time, place, or person associations with the identified cases (see the section on "Make Time, Place, and/or Person Associations").

Review recently received complaints in the water-related complaint log (Form B). Contact other nearby health agencies, hospital emergency rooms, elderly care centers, and local physicians to discover other epidemiologically related cases. Call previously contacted persons to see whether they know anyone else who has become ill or had a common association suggested by data in the log. The illness you are investigating may be part of a larger multijurisdictional outbreak, and therefore communicate with adjoining local and state agencies to learn if they are seeing similar illnesses. State or provincial public health agencies can check reportable disease records and state/provincial public health laboratories can start looking for clusters in isolates that they are characterizing. For outbreaks where intentional contamination of water is suspected or confirmed, public health and law enforcement agency officials may conduct the investigation jointly.

If it becomes apparent that an outbreak is associated with a specific water supply (source) or recreational water or event, use Form D1 for recording information. At this stage of the investigation, interviews can be expedited by reviewing the event itself to stimulate each person's recall. Ask about specific symptoms and signs that are known to be common to the syndrome, as well as, time of ingestion or contact with water and onset of illness. Mention each source of water to which the person may have been exposed, and ask each person (whether a case and well persons at risk) which of the water sources had been ingested or contacted.

The number of persons to be interviewed depends on the number exposed and the proportion of them who are probably affected; if fewer than 100 persons were at risk, try to interview all of them; if several hundred are involved, interview a representative sample. Be sure to obtain clinical specimens from these cases and well persons at risk (controls). It is more difficult to obtain positive results if symptoms from persons have ceased. There may be situations where self-administered questionnaires are sent to cases and persons at risk. Use either Form C or Form D or modified versions for this purpose. After questionnaires have been completed, summarize the data on Form D. Also, identify and interview secondary cases if they become apparent.

Because no two waterborne disease outbreaks are identical, the order of the expanded investigation may not always follow the outlined sequence of procedures. Some investigative steps can usually be done simultaneously by different investigators. Additional procedures may also be required. The principles and techniques described will suffice for most investigations. Modify forms, if necessary, to accommodate the type and amount of information to be collected.

Sources and Modes of Contamination and Ways by Which the Contaminants Survived Treatment

Make on-site observations. Prove or refute hypotheses developed during the epidemiological portion of the investigation. Focus on sources and modes of contamination and ways contaminants could survive and pass through water treatment. As applicable, conduct an on-site investigation of source (lakes, streams, areas around groundwater, etc.), treatment facilities, distribution lines, cross connections, water reservoirs, places of recreational water contact and/or sites at which aerosols were generated. Such an epidemiologically focused investigation is quite different from sanitary surveys done during routine evaluations of water source sites, treatment plants or recreational water facilities.

Not all drinking water (even municipal and bottled water) is disinfected; so, it is important to identify whether the water source is treated and if so, how. Some treatments (filtration, reverse osmosis, membrane treatments, riverbank filtration, and others) may not be complemented with a disinfection step. Sanitary survey information can provide information about potential sources of contamination in the area of a usually pristine water source. Microbiological records of a water supplier, particularly if any total coliform positive samples were found by the system in the last 6 months, may help identify a contamination pathway. If significant matters relating to water quality are observed or otherwise identified during the investigation, note them and communicate them to those responsible for the water system and to the proper authorities. Do not lose the focus and objectivity of the investigation by confusing matters of quality and aesthetics with factors related to contamination by, and survival of, infectious and toxic agents. Use the HACCP-system, also known as systems analysis, way of thinking in your investigation.

Plan On-Site investigation

Contact the person with the highest responsibility for the operation and maintenance for the implicated water source, water treatment facility, and/or distribution
lines. Identify the types of records that ought to be reviewed during the investigation and their likely source. Do not forget that the responsible authorities also can
have records (about water quality, if there has been a change of municipal water
supply, industrial water pollution, wastewater pollution). They can be good sources
of information about recent pipe breaks and other water system issues that could be
related. In many cases they will be aware of the potential for contamination upstream
of source water intakes. If applicable, obtain water distribution maps and recent
water quality reports from appropriate departments. If you are not familiar with the
community in which the investigation is to be done, obtain maps of the area to
locate streams, lakes, water treatment facilities, and other community features that
might have a bearing on the investigation. Check if there are water protection areas
and their rules. Get plans and specifications on design of treatment facilities from
consulting engineers or state agencies that approve these facilities. Contact weather
bureaus, airports, radio/television stations, or newspapers for information on heavy
rainfall, flooding, extremely low temperatures, droughts, or other unusual weather
conditions that preceded the outbreak, if this information is unknown to investigators. Contact police or fire departments about traffic accidents, which can be the
source of the outbreak. Review all background data pertaining to the suspect water.
As information is gathered, record it on applicable parts of Form G.

Discuss with laboratory personnel that a field investigation will be made, and get
their suggestions regarding samples and specimens that should be collected (see
Tables E and F). Confer with them about special analyses, media, and sampling
procedures; make arrangements for rapid transport of samples to the laboratory. The
samples must maybe be transported at the right temperature. Pick up appropriate
forms and sample collection equipment (preferably preassembled in a kit—see
Table A). The laboratory can probably help assemble this kit.

Identify Contributory Factors of Outbreaks

During the investigation, identify factors that contributed to contamination and survival of the etiologic agents and perhaps also to their growth or amplification or
another cause of the outbreak. Identified factors and situations that have contributed
to waterborne disease outbreaks include those listed in Table 2.

Focus the investigation on the potential situations listed in Table 2, as applicable.
Remember that other possibilities can occur. Describe circumstances that contributed
to contamination and that permitted the etiologic agent to survive so that it reached
drinking, agricultural, industrial, or recreational water. Also describe circumstances
that allowed pathogenic bacteria or algae to multiply in the water. Write your findings
down on the back of Form G1 (Illustration of contamination flow) or on a separate
sheet. Continually update the listing in Table 2 with newly available data.

Table 2 Factors that have contributed to waterborne disease outbreaks according to various water sources and the following systems

Source/system	Factors
Surface water	Ingestion of untreated surface water
	Contamination of watershed by human or animal feces
	Use of contaminated surface water for supplementary water source
	Water from sewage treatment facilities
	Overflow of sewage or outfalls near water intake
	Heavy rains and/or flooding
	Contamination from algal blooming
	Dead animals in stream or reservoir
	Live animals and birds in stream, reservoir or watershed
	Poorer quality of water supply for economic reasons
	Accidental industrial pollution of water
	Traffic accidents with transportation of chemicals
	Fire drill sites—fire foam
Groundwater	Overflow or seepage of sewage into well or spring
	Surface runoff into well or spring
	Contamination through limestone or fissured rock
	Heavy rains and/or flooding
	Contamination by pesticides or other chemicals
	Seepage from abandoned well
	Contamination of raw-water transmission line or suction pipe
	Improper well construction and lack of maintenance
	Surface water percolation
	Migrating landfill leachates
	Contamination from grazing animals and from their manure
	Pests (e.g., rodents, snakes) can come into well
Inadequate treatment of water or other problems in facilities	No disinfection or too much disinfection
	Inadequate concentration or contact time of disinfectant
	Interruption of disinfection
	Leakage of sewage water to the drinking water (e.g., from floor drains)
	Inadvertent by-pass of treatment process
	UV-light treatment not functioning (improper cleaning and maintenance of lamps/bulbs)
	No functioning of alarm system
	Inadequate filtration
	Inadequate prefiltration treatment
	Excessive fluoridation
	Excessive dosage of process chemicals
Storage/transportation deficiencies	Unprotected storage tanks, reservoirs, pumping stations, reservoirs, hydrants or tanks

(continued)

Table 2 (continued)

Source/system	Factors
	Contamination of cistern or individual storage facility by surface water runoff, sewage seepage or nearby animal clustering (e.g., feed lots)
	Leakage of sewage water to the drinking water
	Improper or no disinfection of new storage facility
	Lack of maintenance
	Microbial growth in water reservoirs
	Microbial after growth in pipes and tanks
	Unsuitable material in contact with water
Distribution/plumbing deficiencies	Back siphonage
	Cross connections
	Illegal connections
	Corrosion inhibitors not added when the water supply is known to have industrial chemicals
	Contamination of mains during construction or repair
	Water main and sewer in same trench or inadequately separated or inadequately overpressure
	Improper or no disinfection of mains or plumbing
	Unaccounted water loss
	Unauthorized tap-ins
	Frequent line breaks
	Lead pipes not replaced, especially where the water has a low pH
Other factors from ingestion of water or ice	Use of water not intended for drinking
	Contaminated buckets and other containers
	Drinking water bottles used for chemicals (unlabeled)
	Contaminated drinking fountains
	Contaminated taps
	Deliberate contamination/vandalism
	Contaminated ice
	Hand scooping of ice
Water contact	Puncture injuries or wounds
	Swimming or wading in parasite-infested waters
	Swimming water with contaminations from animals
	Snails in water
	Algal blooms in swimming water
	Sewage contamination of swimming water
	Improper pH adjustment
	Improper chlorination or other disinfectants
	Excessive process chemicals
	Improper filtration
	Rough pool wall construction
Aerosolized water	Stagnation of water
	Water temperatures conducive to growth of pathogen

(continued)

Table 2 (continued)

Source/system	Factors
	Dead ends in water distribution lines
	Generation of aerosols
	Poorly maintained air conditioning units
	Poorly maintained humidifier for fruits and vegetables in grocery stores
	Poorly operated and/or maintained water systems
	Contaminated cooling towers during sustained heat spells that tax air-conditioning systems
	Excessive exposure to showers, running faucets, waterfalls, irrigation and misting systems

Meet Managers

Introduce yourself (who you are, where you come from, who ordered you there) to the owner, resident, or persons in charge and state your purpose, when you arrive at the place of the suspected contaminated water source. Emphasize that your visit is to confirm or eliminate suspicion that this water was a source of illness. Tell him or her that a complete epidemiologic study is in progress and that other possible sources (such as food) will be investigated as well as operations of this site. Explain that your investigation is not to fix blame but to identify the cause of the outbreak. Emphasize that the findings can yield benefits related to the ability to identify needed improvements, to educate staff and to provide public support. Try to create an atmosphere of cooperation. Maintain an open mind and try to answer all questions. If you can not answer a question, tell the person that you will come back with an answer. Come back to the person within 1 or 2 weeks even if you do not have any new information. Give the person your phone number and e-mail address and tell the person to contact you if the person has more information later.

Privately interview key persons responsible for operating or repairing water facilities. Do not forget to interview persons from other work shifts. Identify persons who were working there at the likely time of contamination and have since left and interview them as well. Ask questions to determine the flow of water and operations from intake through distribution through plumbing systems. Ask about any changes in operation, unusual events in the watershed or repairs to the water facilities. Ask if you can check records, both in paper form and on the computer (monitoring system), analyses of results, and/or incident reports.

Plant personnel may not describe water treatment or installations exactly as they existed at the time that a mishap occurred. They may fear criticism or punitive action as a result of their possible role in the causation of the outbreak. Their descriptions should be plausible and should account for possible sources and modes of contamination and indicate possibilities for survival of pathogens. If a description does not contain all the information desired, reword questions and continue the inquiry. Confirm accounts by private interviews with others knowledgeable of water treatment or operation of the facility. Be alert for inconsistencies among the accounts told by different persons.

Seek resolution of discrepancies in accounts by watching actual procedures as they are being carried out, by taking appropriate samples, or by conducting experiments. A communicative working relationship between the plant management and the investigator influences plant workers' attitudes toward the investigative team. Consider the position, feelings, and concerns of the manager and staff; defensive reactions are normal on their part.

Diagram each phase of the water system or situation under study on Form G1 (Illustration of contamination flow). Insert special symbols and notes for all sites that might be involved in introducing contamination to the water or where contaminants might have survived treatment. Record other information gathered on the appropriate parts of Forms G2–8.

Gather and Review Records

Review and collate appropriate information on quality control and operational records from the water utility and from responsible agencies. As applicable, obtain information on quality of untreated surface or groundwater from a local, state/ provincial, or national pollution control or geological survey agency. Also, seek water distribution maps, well logs, descriptions of geological conditions and indices of groundwater quality from them. For surface water supplies, obtain information on upstream discharges and unusual events that may have affected raw water quality.

Get data on finished water quality in the distribution system from a local, state/ provincial water surveillance or regulatory agency. Water suppliers also frequently have records of raw and finished water quality. Review data on quality control tests (e.g., pH, chlorine residual, chlorine demand, bacteriological and chemical tests, turbidity, jar test data, incident reports) that are available. Obtain data on cross connection control programs and sewer repairs from the water supplier or other local agencies (e.g., building inspectors, sewage departments). Review files for data concerning potential sources of contamination for individual or semipublic water supplies (e.g., diagrams of septic tank systems, sewer line locations, well logs, small individual wastewater plants, accidental industrial pollution of water, traffic accidents involving chemicals, salting of roads or sawmills).

Check if they have any HACCP-systems or water safety plan and, if so, how they monitor their CCPs (critical control points) and if they are implementing control measures. Ask them about HACCP, to see if they understand the system and if results are documented. Check if the HACCP-system is validated (should be documented) and that they are conducting internal audits.

Get information on all aspects of normal operations as well as unusual events or conditions to determine whether such events were coincidental with the time of suspected contamination as determined from the epidemic curve. Also, consider the time it takes for a contaminant in the raw water or treatment plant to reach households in the affected community. Ask responsible persons for this information.

Compare data on heterotrophic plate and total coliform counts of raw and finished water leaving the treatment facility and of water in distribution lines. Also, compare data on chlorine residuals within the plant with that in distribution and check, if they have, that the UV-light is functioning. Review other test data (e.g., turbidity and chemical analyses) that may indicate a problem situation. Identify locations and dates of sample collection. Take photos, if it is allowed, of things you suspect are not right. Go back more than one incubation period of the disease under investigation. Record this information on Form G2 (Record review of on-site investigations, and test results prior to and during outbreak). Photocopy appropriate records for confirmation and subsequent review and attach them to the record review form. Be alert for evidence of falsification of records. While reviewing records, watch for evidence of the following:

- Potential contamination of groundwater systems because of proximity to septic tank systems, latrines, animals manure or landfills, industrial contamination of the water supply, small sewage plants, especially old and nearly forgotten ones, and recent heavy rain
- High heterotrophic plate or coliform counts, or counts that exceeded the average (median) or typical count
- Sudden changes in water quality or operating practices that suggest the possibility of contamination or treatment failure
- High turbidity, unusual odor, color, or taste, or high coliform counts in raw water, which can indicate potential overloading of the normal treatment process
- High levels of ammonium, nitrate and nitrite, which can indicate organic and inorganic contaminants
- Low chlorine residuals in treated water or higher-than-normal amounts of chlorine used, which can indicate either a high chlorine demand or a sudden high level of contamination
- A sudden drop in amount of disinfectant used, possibly indicating failure or interruption of a disinfection process. No functioning alarm system
- A sudden change in the amount of a chemical (e.g., alum or ferric sulfate) used, suggesting equipment disfunction or inadequate coagulation or flocculation and thus poor filtration
- Lack of treatment chemicals if a more corrosive water supply is used (see Box 3, The Flint Water Crisis)
- Pump failures, draining of distribution lines or reservoirs, or massive pumping to fight fires, which can produce low pressures that can cause contamination through cross connections or back siphonage
- Repairs to water mains, wells or pumps where contamination could have been introduced

Record this information on Form G2 or other appropriate form in the G series.

Conduct On-Site Investigations

As applicable, investigate the water source, treatment facility, distribution and plumbing systems, sites where water was contacted, and sites at which microorganisms amplified and aerosols disseminated. Use forms in the G series as guides while observing facilities, gathering data, making measurements and collecting samples. Google or Bing maps or other similar resources' views of the watershed can be very helpful in identifying potential sources of contamination that you will need to investigate further. These maps can also facilitate your own map and diagram making on Form G1.

Investigate the Water Source

The water source may be surface or ground or in some cases a combination of the two. Verify this by observations at the site and by talking to the property owner or persons responsible for operation or maintenance of water supply or recreational facilities, as applicable. Examination of "weather events" such as heavy rainfall may indicate a potential for surface water contamination (See Box 2, The Walkerton Outbreak).

When a surface water is either suspected or implicated as the source of a contaminated supply, get information about the watershed concerning possible sites of contamination of the suspected etiologic agent. This includes, but is not limited to (a) land use, (b) sewage effluent from treatment plants and septic tanks, (c) industrial plants that may be discharging toxic waste, (d) mining wastes, (e) landfill leachates, (f) slaughterhouse discharge wastes, (g) animal feed lots, (h) both domestic and wild animals that use the source water for drinking, (i) sludge disposal from sewage treatment plants or septic tanks (e.g., land spreading or lagoons), (j) storm water discharge. If this information is not available from records or persons familiar with the site, visit the site and observe possible sources of contamination and pollution (e.g., while traveling by foot, vehicle, boat, or helicopter, as applicable). Record this information on Form G3 (Source and mode of contamination of surface water). Diagram the surface water and sites of contamination on Form G1. Note type and location of sources of contamination and their distances from the water.

Visit groundwater sources. Using Form G4 (Source and mode of contamination of groundwaters) as a guide, question owner or operator and inspect groundwater installations to ascertain character of the land and surface and subsurface soil and water. When a well or improved spring is under consideration as the source of the contaminated supply, observe its location relative to possible sites of contamination and to whether its construction allows contaminants to reach the water. Determine locations of all sewage outflows or disposal sites (e.g., septic tanks and absorption lines, cesspools, privies, and other sewage disposal facilities), gradients, and distances from the well or spring. Determine the type of soil at the site. If the soil is limestone or fissured rock or if there is a high ground or perched water table, pollution may travel many miles. In this case, the search for sources of contamination may have to be expanded for a considerable distance from the well or spring. Ascertain whether there were heavy rains, heavy snow melts, or sudden discharge

Box 2 The Walkerton Outbreak

In May, 2000, many people in Walkerton, a small Ontario, Canada, community of about 5,000 people, began to simultaneously experience bloody diarrhea and other gastrointestinal infections. On May 8–12, torrential rain had unknowingly contaminated the town's water system, but operators failed to check residual levels for a period of several days, allowing unchlorinated water to enter the distribution system. However, the privately-owned Walkerton Public Utilities Commission insisted there was no problem with the water despite other laboratory tests showing evidence of *E. coli* contamination. Illnesses began about May 18, with the first death occurring on May 22 and the seventh and last on May 30. By May 21, however, many more cases had been diagnosed, the infectious agent determined to be *E. coli* O157:H7, and contaminated well water was confirmed as the source of the *E. coli*; all this allowed the region's Medical Officer of Health to issue a boil water advisory, warning residents not to drink the tap water. Two days later, laboratory results identified the presence of *Campylobacter* and *E. coli* O151:H7 and DNA testing showed that the contaminating source was a cattle farm a short distance from a well used for the water supply. By the time the outbreak was over, >2300 were ill and 7 had died. The people who died directly from drinking the *E. coli*-contaminated water might have been saved if the Walkerton Public Utilities Commission had admitted to contaminated water sooner. Those in charge of the water utilities at the Commission had no formal training in their positions, retaining their jobs through three decades of on-the-job experience. They were later found to fail to use adequate doses of chlorine, fail to monitor chlorine residuals daily, make false entries about residuals in daily operating records, and misstate the locations at which microbiological samples were taken. Regulations state that water suppliers are required to treat groundwater with chlorine to sufficiently neutralize contaminants and sustain a chlorine residual of 0.5 mg/L of water after 15 min of contact. Had utility operators adhered to the protocol, the disaster most likely would have been averted. The operators knew that these practices were unacceptable and contrary to Ministry of Environment guidelines and directives; they eventually admitted falsifying reports and were sentenced to short jail terms. The Ontario government was also blamed for not regulating water quality and not enforcing the guidelines that had been in place. The water testing had been privatized in October 1996. An enquiry found that the water supply, drawn from groundwater, became contaminated with the *E. coli* O157:H7 strain from manure from cattle on a farm washing into a shallow water supply well after heavy rainfall. The risk of contamination from farm runoff into the adjacent water well had been known since 1978. Key recommendations from the enquiry included source water protection as part of a comprehensive multi-barrier approach, the training and certification of operators, a quality management system for water suppliers, and more competent enforcement, which were incorporated into Ontario new legislation. The bottom line of the enquiry was that officials and municipal water facilities operators and managers across North America need to recognize public waters are a most valued but vulnerable public resource. Investment in keeping them safe and secure needs to be a community top priority.

from dams that could have resulted in flooding within the duration of the incubation period of the disease under investigation.

Obtain information on the depth of the well in reference to the ground water table from the owner or by referring to any available well logs on public file or from local drillers. Observe well construction and get information about depth of casing, depth and method of grouting, and whether there is an underground discharge. Observe whether there is an impervious well platform and whether the pump or casing seal was subjected to flooding. Illustrate the situation by showing location of the well in reference to possible sites of contamination on Form G1. Note distances between the well and contamination sites and elevations. Determine whether any pumps were out of order or had been repaired during the interval of concern. If priming of the pump was done, find out the source of the water used. Record this information on Form G4. Test hypotheses of modes of contamination by conducting a dye test and/or sampling the water. (See appropriate sections of this manual for directions.)

Collect samples of water from these sites and submit them for analysis of the suspected etiologic agent or for any physical, bacterial, or chemical tests that will provide evidence of contamination or movement of the contaminants. (See Procedures for collecting water samples) Record these results on Forms G3 or G4 and I (Laboratory Results Summary). When appropriate, confirm hypotheses by a dye or other tracer test. (See section on this subject).

Investigate the Water Treatment Facility or Individual Treatment Devices

Determine the means by which the etiologic agent survived treatment or was otherwise not eliminated or inactivated. Consider the treatment process as a series of barriers placed between contaminants and consumption of the treated water. The operation of each barrier should be optimized. Review available data for each step in the treatment process. Records of well-maintained and properly calibrated continuous monitoring equipment will be especially valuable. Look for failures in the barriers, which could include (a) lack of disinfection, (b) inadequate concentration of disinfectant or contact time, (c) interruption of disinfection, (d) inadequate filtration, (e) lack of corrosion inhibitors, which may follow inadequate pre-filtration treatments. In 2015 in Flint, Michigan excessive levels of lead were found in drinking water from corrosion of water distribution pipes (see Box 3, The Flint Water Crisis). Corrosion inhibitor had not been added. Also, look for possible introduction of contaminants within the treatment process, such as in treatment chemicals.

Box 3 The Flint Water Crisis

This event is considered a disaster, still unfolding, initiated from a political decision to save money, and ending up with acute and chronic illnesses and deaths to residents of a Michigan city, as well as high system remediation and health-related costs to the taxpayer. On April 24, 2014, Flint, Genesee County, Michigan, switched its water supply from Detroit's system to the Flint River as a cost-saving measure for the struggling, majority-black city on the recommendation of the state-appointed emergency manager. Flint agreed to separate from the Detroit Water and Sewerage Department and go with the Karegnondi Water Authority, including the decision to pump Flint River water. This was to be an interim measure until a new pipeline from Lake Huron was constructed in 2016 to serve the region. Soon after the switch, residents begin to complain about the water's color, taste and odor, and to report rashes and concerns about bacteria. In August and September 2014 city officials issued boil-water advisories after coliform bacteria were detected in tap water. In October 2014, the Michigan Department of Environmental Quality (MDEQ) blamed the cold weather, aging pipes and a population decline. In the same month a General Motors plant in Flint

(continued)

stopped using municipal water, saying it was rusting car parts. On January 4, 2015, the city announced that Flint's water contained a high level of trihalomethanes, a byproduct from increased disinfection by the city. Though this is in violation of the Safe Drinking Water Act, officials told residents with normal immune systems that they have nothing to worry about. In January 2015, Detroit's water system offered to reconnect to Flint, waiving a $4 million connection fee but the offer was declined by the emergency manager. By February, State officials continued to play down any water problems saying that the water was not an imminent "threat to public health." On February 18, 104 parts per billion (ppb) of lead were detected in drinking water at a Flint home and the federal Environmental Protection Agency (EPA) was notified. The EPA does not require action until levels reach 15 parts per billion, but science indicates that there is no safe level for lead in potable water. Officials from EPA and MDEQ discussed the lead level in the sample, and EPA found that the State was testing the water in a way that could profoundly understate the lead levels. On March 3, 2015, a second testing detected 397 ppb of lead in Flint drinking water. A consultant group hired by Flint, reported that the city's water met state and federal standards, and it did not specifically report on any lead levels. In May, tests revealed high lead levels in two more homes in Flint. In July, an EPA administrator told Flint's mayor that "it would be premature to draw any conclusions" based on a leaked internal EPA memo regarding lead. However, in September, Flint was asked to stop using the Flint water supply or consider corrosion control for it, because it was causing lead to leach from the water pipes and children had high levels in their blood. State regulators insisted the water was safe. Nevertheless, on October 1, the Governor of Michigan ordered the distribution of filters, the testing of water in schools, and the expansion of water and blood testing after a briefing on the lead problems with the MDEQ and federal officials. At the same time, Flint city officials urged residents to stop drinking water. On October 16, Flint reconnected to Detroit's water system, and residents were advised not to use unfiltered tap water for drinking, cooking or bathing. On October 19, the Director of MDEQ reported that his staff had used inappropriate federal protocol for corrosion control, and soon after, the Governor announced that an independent advisory task force would review water use and testing in Flint. On December 9 Flint added additional corrosion controls, and soon after an emergency was declared. At the end of December, the task force found that the MDEQ was accountable for its lack of appropriate action, and the Director resigned. On January 16, 2016, the Governor asked the National Guard to distribute bottled water and filters in Flint, and President Obama declared a state of emergency in the city and surrounding county, allowing the Federal Emergency Management Agency to provide up to $5 million in aid.

Three days earlier the crisis expanded to include Legionnaires' disease, because of a spike in cases, including ten deaths, after the city started using river water. On January 21, the Michigan Department of Health and Human Services stated it did not have enough information to conclude that the increase in cases was related to the ongoing Flint water crisis, although the. head of Michigan's Communicable Disease Division had said three months earlier that the number of *Legionella* cases at that time "likely represents the tip of the iceberg." As of February 2016, the number of reported cases was close to 100. A Flint hospital official was surprised that Michigan and local health agencies did not inform the public about the Legionnaires' outbreak in Genesee County in 2014–15 until January 13; the hospital earlier had spent more than $300,000 on a water treatment system and bought bottled water for patients. The source of *Legionella* is not known but it was likely in the Flint River, and possibly extensive flushing of Flint's colored water, which had undesirable odors and tastes, by residents may have caused chlorine residual in the pipes to be washed away, leaving the pipes susceptible to growth of the *Legionella*; in addition, aerosols from the extensive flushing from turned-on faucets might have led to close contact between the residents and the pathogen. The investigation of the cause of the illnesses continues with criminal charges laid against Michigan departmental employees.

Observe treatment processes from the water inlet to the finished water discharge. Diagram on Form G1 the treatment process; insert notations of hazardous situations that were observed. Collect samples of water at the inlet, after each phase of treatment that may have functioned suboptimally or failed, and at the outlet. Test the samples for pathogens that cause a syndrome characteristic of that being investigated, for indicator organisms and for physical and chemical characteristics of the water, as appropriate to the situation.

Evaluate effectiveness of the disinfection process and resulting residuals. Determine the type of disinfectant (e.g., gaseous chlorine, hypochlorite, chlorine dioxide, chloramine, ozone, ultraviolet irradiation) used and whether the disinfection treatment was adequate for the volume of water treated. Determine, by talking to water treatment plant employees and reviewing records of the plant or regulatory agency, whether there were any interruptions of disinfection during the two weeks prior to the first onset date. Determine contact time between the point of addition of the disinfectant and the first point of use. Measure the chlorine residual, pH and temperature of the water just before it leaves the plant. Observe the condition, operation, and maintenance of disinfectant dispensing equipment. Review plant records to identify any sudden changes in disinfectant demand that causes temporary depletion of disinfectant residuals and allows survival of pathogens. Review maintenance records for disinfectant dispensing equipment and quality assurance records for online analyzers. Record this information on Form G5a (Disinfection failures that allowed survival of pathogens or toxic substances).

Calculate disinfectant rate applied and usage (see formulae in Form 5a). For example, to calculate disinfectant rate, if the flow rate is 1,000,000 gal/day and the dosage is 15 lb/day:

$$15lb/day \div 1,000,000gal/day = 0.000015lb/gal$$
$$0.000015lb/gal \times 454,000mg/lb \times 0.264gal/L =$$
$$1.8mg/L \text{ disinfectant rate applied}$$

The destruction of pathogens is dependent on (a) type and condition of microorganisms present, (b) type of disinfectant used, (c) concentration of available chlorine or other disinfectant, (d) contact time, (e) water temperature, (f) pH, (g) degree of mixing, (h) presence of interfering substances (which may be related to turbidity). Utilize treatment records that provide small scale time resolution, such as online monitoring data, to determine whether the process was stable during the time period in question. Daily averages may provide evidence of massive failures, but will not provide information about whether consistent treatment was being provided.

In general, the relative effectiveness of microorganisms' resistance to free chlorine, from high resistance to low resistance, is as follows:

• Protozoan oocysts (i.e., *Cryptosporidium*)
• Protozoan cysts (i.e., *Giardia, Entamoeba histolytica*)
• Viruses (hepatitis A virus, poliovirus)
• Vegetative bacterial cells (*Shigella, Escherichia coli*)

Protozoan oocysts are highly resistant to chemical disinfectants, but not to physical means such as UV light or ozone (gas). Microorganisms within each group and strains among the same species differ somewhat in resistance. The state of injury induced by environmental impacts and selection of resistant strains influence survival. Aggregation of microorganisms and/or close association with debris shield them to various degrees from lethal effects of disinfectants and attachment to surfaces such as pipe walls to form biofilms that protect organisms from inactivation by disinfectants.

A measurement of microbiological inactivation by disinfectants is the *CT* value (CT_{calc}), which is the product of the free residual disinfectant concentration (*C*) in mg/L that is determined before or at the first user (customer) and the corresponding disinfectant contact time (*T*) in minutes (i.e., $C \times T$). Refer to Table I ($CT_{99.9}$ values for inactivation of *Giardia* cysts at different concentrations of disinfectants, temperatures, and pH values) and Table I (*CT* values for inactivation of viruses at pH 6–9, at different temperatures with different disinfectants for comparing disinfectant efficiencies). Make residual measurements during peak hourly flow. For comparisons of *CT* values between the indicated pH, temperature, and concentration values, use linear interpolation. (For example, for free chlorine, 10°C, concentration 1 mg/L, pH 7.5=[166−112=54; 54/2=27; 112+27]=139). If no interpolation is done, use the $CT_{99.9}$ value at the higher temperature, at the higher pH and higher concentration.

A simple *CT* calculation, for example, using a disinfectant concentration (*C*) at the basin effluent of 1.3 mg/L and a detention time (*T*) of 22 min, is as follows:

$$C \times T = 1.3 \text{mg} / L \times 22 \text{min} = 28.6 \text{mg min}/ L$$

Use this calculation for comparing to values in Table I or J. The calculated *CT* value should be higher than the value stated in the table for specific conditions of disinfection, temperature, pH, and concentration (residual). In this situation, if the temperature of the water was 15°C, the pH 7 and the concentration of chlorine 1 mg/L, a *CT* value of 75 would be needed for a 99.9 reduction of *Giardia* cysts. The *CT* value of 28.6 would have been inadequate to meet the criteria and could explain the survival of the pathogen under investigation.

Microbial inactivation efficiencies vary considerably among different disinfectants and are influenced by the characteristics of the water and water temperature. Tables I and J show that, in general, ozone is more effective than chlorine dioxide, which is more effective than free chlorine, which is more effective than chloramines. Also, in general, longer contact time increases the degree of inactivation, and higher water temperatures as well as lower pH values increase rates of inactivation. Rapid mixing of the disinfectant with water increases disinfection efficiencies, whereas dissolved organic matter reacts with and consumes the disinfectant and forms products that have weak or no disinfection activity. Certain inorganic compounds and particulate matter also react with disinfectants.

The *CT* value must be determined sequentially whenever a disinfectant is added to water. Contact time (*T*) is the duration in minutes for water to move from the point of application of the disinfectant or the previous point of residual disinfectant measurement to the point where residual disinfectant concentration (C) is measured. It is measured from the first point of disinfectant application and from all subsequent applications until or before the water reaches the first user. Determine contact time in pipelines by dividing internal volume of the pipe by the maximal hourly flow rate through the pipe. Determine the flow rate from (a) plant records, (b) continuous monitoring device readings, (c) measurements at hourly intervals, or (d) if this sort of information is unavailable, measurements at expected high flow periods. Use tracer studies to determine contact time within mixing basins and storage reservoirs. These values represent only 90% effectiveness because of short circuiting. Chlorine, fluoride, and rhodamine WT (but not B) are commonly used tracer chemicals. Contact time is usually measured by a step-dose method, but a slug-dose method is used where chemical feed equipment is not available at the designated point of addition or where such equipment does not have the capacity to provide the necessary concentration. (See appropriate EPA literature for procedures, and consider getting engineering expertise if these matters are too complex.)

Estimate whether pathogens had been inactivated. To do this, divide the CT_{calc} value by a value ($CT_{x\%}$) resulting in a certain percentage inactivation (e.g., 99.9% [3-log] or $CT_{99.9}$ for *Giardia* cysts and 99.99% [4-log] or $CT_{99.99}$ for viruses). This gives an inactivation ratio. See Table I for $CT_{99.9}$ values for *Giardia* and Table J for $CT_{99.99}$ values for viruses.

Following is a sample calculation for data in Table I when water temperature is 20°C, pH in a clearwell (reservoir for storing filtered water) is 7.0, time (T) (either calculated or measured by dye test) is 38 min, and the disinfectant used is chlorine: The desired CT value for 99.9% inactivation of *Giardia* for pH 7 at 20°C is between 52 and 68 depending on concentration of disinfectant. In this case, the disinfectant measured at the clearwell outlet is 2.0 mg/L. Therefore,

$$CT \text{ is } 38\,\text{min} \times 2.0\,\text{mg}/\text{L} = 76\,(\text{mg}\,\text{min}/\text{L}).$$

The result, 76, is larger than the value, 62, required in the table; hence, these disinfection concentration (C) and time (T) conditions should result in a 99.9% or greater inactivation of *Giardia* cysts. For free chlorine, a 3-log inactivation of *Giardia* cysts provides greater than a 4-log inactivation of viruses.

The following example, using the data in Table I, demonstrates a means to calculate the increased disinfectant dosage needed for a plant during the transition from summer to winter, when the water temperature fell from 15 to 5°C, chlorine dioxide was the disinfectant used and the T value (calculated or measured) is 12 min.

Using Table I, the required CT at 15°C for a 3-log inactivation of *Giardia* cysts by chlorine dioxide is 19. Therefore,

$$19\,\text{mg}\,\text{min}/\text{L} \div 12\,\text{min} = 1.58\,\text{mg}/\text{L}.$$

The CT value for 5°C for this disinfectant for the same inactivation is 26. Therefore,

$$26\,\text{mg}\,\text{min}/\text{L} \div 12\,\text{min} = 2.17\,\text{mg}/\text{L}.$$

Table 3 Sum of calculated CT values for free chlorine

Location	Disinfection (mg/L)	Contact (min)	$C \times T$ (CT_{calc})	pH	Water temperature (°C)	$CT_{99.9}$ (Table I)	$CT_{calc}/CT_{99.9}$
Basin 1	1.3	30	39	7	15	76	0.513
Basin 2	1.0	25	25	8	15	108	0.231
Basin 3	0.8	60	48	8	15	105	0.457
SUM							1.201

In this situation, the plant should have increased the chlorine dioxide concentration from 1.58 mg/L to 2.17 mg/L to maintain the same efficiency of disinfection. If this had not been done, it may explain the survival of pathogens in the water supply.

The sum of these ratios gives the total inactivation ratio, which should equal 1 or more to provide effective disinfection. Make calculations and record information on Form G5a. The following example shows the way this is done. Chlorine is added to three basins. Chlorine concentration, contact time, temperature and pH are measured at these locations and recorded as shown in Table 3. Data from Table I is combined to do the calculation.

The resulting sum exceeds 1.0. This ensures that the plant met the recommended or required *CT*.

Regulations may require that community and non-community public water systems that use surface water or water under direct influence of surface water meet a criterion (e.g., minimum of 99.9% [3-log] removal and/or inactivation of *Giardia* cysts and a minimum of 99.99% [4-log] removal and/or inactivation of viruses of fecal origin that are infectious to humans). Removal and/or inactivation of microorganisms may be accomplished by either filtration plus disinfection or disinfection alone, depending on the water source. Water systems using chlorine with *CT* values that attain minimal level or inactivation of *Giardia* cysts will result in inactivation of 99.99% (4-log) of viruses.

Evaluate the prefiltration processes (e.g., coagulation, flocculation and sedimentation). Coagulation is a process that uses coagulant chemicals and mixing, by which colloidal and suspended materials are destabilized and aggregated into flocs. Flocculation is the process that enhances agglutination of smaller floc particles into larger ones by stirring. Sedimentation is the process by which solids are removed by gravity separation before filtration. Observe whether these processes reduce turbidity. Calculate detention (transit) time within the settling tank and seek information about frequency and method of cleaning the tank. For example, if an 8-ft-deep sedimentation basin has a volume of 1 million gal, and the plant flow rate is 20 million gal/day, detention time in the basin is: (in your country you may want to calculate rates based on metric measurements)

$$1,000,000 \text{gal} \div 20,000,000 \text{gal}/\text{day} = 0.05 \text{days} (\text{or} 1.2\text{h, or } 72\text{min.})$$
$$\text{Then, depth} / \text{time is} :$$

$$8\text{ft} \div 72\text{min} = 0.11\text{ft} / \text{min}.$$

Several different types of filtration may be used in water treatment facilities. These are conventional, direct (both conventional and direct are referred to as "rapid" filtration), slow sand filtration, and diatomaceous earth filtration. Conventional filtration consists of a series of processes including coagulation, flocculation, sedimentation, and filtration. Direct filtration consists of a series of processes including coagulation and filtration but excluding sedimentation. Slow sand filtration is a process involving passage of raw water through a bed of sand at low velocity (usually less than 0.4 m/h), utilizing both physical and biological means to remove particles and microorganisms. In diatomaceous earth filtration, water is passed through a precoat cake of diatomaceous earth filter medium while additional filter medium is continuously added to the feed water to maintain the permeability of the filter cake.

If done properly, each filtration method results in substantial particulate removal. When rapid sand filters have a head loss of about 7–10 ft, they require back washing. Filters are backwashed by reversing the flow of the filtered water back through the filter at a rate between 15 and 30 gal/min/ft² of sand-bed area. Sometimes water jets at the surface aid in loosening and removing deposited material on the sand. Observe an actual backwash and look for indications of short-circuiting or areas of the filter material that seem agglomerated or resist being cleaned by the flowing water. If backwash water is not discharged to waste, evaluate where it is released. Slow sand filters eventually become clogged. When this occurs, a scraper or flat shovel is used to remove the top layer of clogged sand, and new sand (equivalent to the depth removed by scraping) replaces the old.

Test the effectiveness of filtration for each filter unit by observing capacity and filtering area relative to volume and turbidity of the filtered water. Also, review turbidity, headloss, and filter rate record. Look for anomalies, especially in the few hours after a filter is returned to service, and before the filter is backwashed. Review criteria that cause a backwash to be initiated, and establish if these criteria were followed during the time preceeding the outbreak. Determine the source of backwash water and the frequency of back washing of filters from records and head gauge readings. Check whether the water used to back wash or clean filters came from an untreated source and determine the fate of the backwash water. In the case of illness due to chemical substances, evaluate types of chemicals used and condition, operation and maintenance of chemical feeding equipment. Consider sampling backwash water for pathogens under investigation. Review plant records for results of monitoring and be alert for changes that suggest treatment failure. Record this information on Form G5b (Source of contamination and treatment failures that allowed survival of pathogens or toxic substances.)

Data in Table K (Estimated removal of *Giardia* and viruses by various methods of filtration), give a summary of expected minimal removal of *Giardia* and viruses in well operated filtration systems. Values can be subtracted from *CT* values required for disinfection.

Although contamination is likely to be associated with raw incoming surface water, look for bypass connections where raw or partly treated water can get into treated water. Also look for common walls that separate treated and untreated water.

Consider the possibility that a contaminant was introduced in any of the treatment chemicals themselves, or as an act of sabotage. Determine whether any flooding has occurred during the interval of concern. Check absentee records for possible enteric illness of the water treatment plant staff. Such illness may reflect either sources of contamination or victims. Record this information on Form G5b.

At domestic locations, evaluate treatment devices (such as chlorinators, filtration units, softening equipment) as described above, but modified to fit the situation under investigation. Record observations and measurements on Forms G5a and G5b, as applicable.

Evaluate the Water Distribution and Plumbing Systems

The water distribution system can be complex, with multiple entrance points for treated water and different pressure zones in which water can enter but not leave. Water flows in the direction in which it is being "requested," so can flow in different directions in the same pipes from one hour to the next. Contaminated water can enter a potable water supply from a non-potable water supply when the two are directly connected. Such interconnections are referred to as cross connections. To evaluate such situations, trace lines of the treated supply from the point of treatment or entrance into a building to points of use and associated plumbing. Look for any interconnections of other water supplies, such as wells, waste lines or holding tanks for water intended for fire control. If cross connections are found, look to see whether backflow prevention devices are inserted between the lines and, if so, whether they are functioning properly. Also, look to see whether there is an air gap between the water inlet and vessel or tank. Evaluate the arrangement and operation of check valves on connections between the two water systems. Review inspection report for backflow prevention devices.

Contaminated water can also enter a treated supply by siphonage from a contaminated vessel or sewerage to the potable water line having negative pressure. This is referred to as back siphonage. Examine all water vessels to see whether they contain submerged inlets or hoses connected to water faucets, and if so, whether properly functioning vacuum breakers are in place. Without proper air gaps or properly functioning vacuum breakers, there is a possibility of siphonage of water from plumbing fixtures in upper stories to lower stories when line pressure is negative. This may occur when faucets on lower floors are opened after the water supply valve has been turned off for repairs or when the supply line has had a sudden loss of pressure, as can happen with nearby heavy use of water (e.g., to fight fires or irrigate) or when pressure lines are broken. Measure water pressure on upper stories of buildings to determine whether negative pressure occurs. (Pressure losses may be transient and of very short duration.)

Interview building managers and residents about whether there were (a) any repairs of water service during the past month, (b) fires that occurred nearby, or (c) other situations that could have caused negative pressure in the water line. Also, if appropriate, review fire and utility department records for information about these situations. Get dates of line repairs to see whether they correlate with the time of incubation periods of early cases. Measure chlorine residual (of chlorinated water systems) and take samples for microbiological tests at several strategic locations in the distribution systems. Perform calculation on comparison of disinfectant residual. If a toxic chemical poisoning is under investigation, talk to home owner, building manager or maintenance staff about whether pesticides or other toxic compounds were sprayed with equipment connected to a hose or a sprinkler system. Furthermore, interview building managers and residents about whether there are persons residing there who either are or recently were ill with diarrhea. They may represent sources of the etiologic agent or may identify victims. Interview those identified about the onset of their illness and symptoms and examine their plumbing systems. Record information obtained during the investigation of distribution and plumbing systems, and record related calculations on Form G6 (Source and mode of contamination during distribution and at point-of-use).

Investigate Water Contact Sites

Evaluate implicated waters used for swimming, water skiing, bathing, clothes washing by hand, or agricultural activities, in a manner similar to that described under the section on investigation of surface water source. If the potential site of contact was natural surface water, determine whether the water is likely to be infested by parasites and look for the presence of snails (Swimmer's itch). For swimming pools, measure the water's pH, chlorine residual, water temperature, and turbidity, if applicable. Also, review pool records for previous information on these characteristics. High turbidity in pools, hot tubs, and spas is a sign of either poor filtration or inadequate disinfection. Evaluate whether the resulting water would adequately protect those who swam or waded in it or had any physical contact with it. Evaluate filter and chlorination equipment as described for water treatment. Backwash filters and collect a sample to get an indication of microorganisms present on the filter (thus obtaining historical information). This approach has been useful for identification of *Pseudomonas aeruginosa*. Look for the presence of slime on tub, whirlpool, slide and pool surfaces, and collect some of this material for analysis for *P. aeruginosa*. If the answer is not obvious, ask ill persons whether they had puncture injuries or wounds or scrapes while immersed in water. Record this information on appropriate parts of Form G7 (Contamination source and survival of pathogens or toxic substances for recreational waters). Collect samples of the water (see section on "Collect Water Samples"), and test them for pathogens and/or indicator organisms, as applicable.

Investigate Sites at Which Respiratory-Acquired Waterborne Agents Amplified and Were Disseminated by Aerosols

The agents listed in Table D can multiply in water and if such water is aerosolized, they can be transmitted to human beings via the respiratory route. Highly susceptible persons (e.g., the elderly, smokers, immunosuppressed individuals) are the usual victims.

Look for possible sites where water may have been or is being disseminated as aerosols. Consider (a) air conditioning cooling towers and evaporative towers, (b) hot water systems (heaters and tanks), (c) shower heads, (d) faucets with aerators, (e) mist machines used to freshen fruits and vegetables in markets, (f) humidifiers, (g) nebulizers/respiratory therapy equipment, (h) whirlpools and spas, (i) dental drills and cleaners, (j) cooling water apparatus for grinders, (k) splash from hoses, (l) water pressure line breaks, (m) decorative water features, (n) outside misters, (o) other aerosol-producing devices. Sample water from all suspect sites for *Legionella* or other waterborne agents that may cause illness when inhaled.

It is not possible to recognize by visual inspection the potential for water to be contaminated with legionellae. Warm temperatures, especially those between 27°C (80°F) and 46°C (115°F), are conducive to growth of legionellae. Additionally, stagnant water allows time for legionellae to multiply, especially in dead-end lines, reservoirs and hot water tanks, and in water trapped in shower heads and faucet aerators. If it is deemed appropriate or necessary to sample for detection of *Legionella* in the environment, collect water samples from suspect sources. It is important to use a lab with proven expertise in isolating and characterizing *Legionella*, such as those labs in the U.S. certified under the Environmental Legionella Isolation Techniques Evaluation (ELITE) Program. The Centers for Disease Control (CDC) have a convenient form for recording case histories (http:// www.cdc.gov/legionella/downloads/case-report-form.pdf).

It is not appropriate to sample air for detection of *Legionella* hazards. It may, however, be appropriate to use micromanometers or smoke to trace direction of air flow to determine route of dissemination. Micromanometers measure pressure differences, and flow can be assumed to travel from high to low pressure areas. Smoke moves from areas of higher pressure to areas of lower pressure and is extremely sensitive to air currents. Observe direction and spread of smoke movement. Record this information on Form G8 (Contamination source and sites of amplification and aerosolization of pathogens).

Collect Water Samples

Prior to the collection of samples, investigators should consult with the testing laboratory that will be used, to receive specific laboratory sampling instructions and sampling kits. Sampling Protocols for potable and non-potable sources are dependent on the specific etiological agent and the related analytical procedures performed by the testing laboratory.

Collect samples promptly to test for possible etiologic agents and for microorganisms indicative of fecal contamination. Contaminants in water are in a dynamic state; their presence and quantity differ with time and place. See Table F (General instructions for collecting drinking water samples) for guidance on collecting and shipping samples for viral, bacterial, and parasitic analyses.

Samples for bacteriological tests can be collected in one of three ways: (a) by letting a stream of water flow into a container or by submersing a container into a volume of water, (b) by passing a large volume of water through a filter, (c) by putting Moore swabs (see Table A for description) or similar absorbent materials in surface water or drains for a few days (see Table F).

Use bottles that have been cleaned, rinsed, and sterilized, or use sterile plastic bags to collect and store samples for bacteriological examination. For a chlorinated water supply, or when in doubt about the presence of residual chlorine, use bottles containing 100 mg/L sodium thiosulfate to combine with any free chlorine in the sample and prevent lethal effects of chlorine on microorganisms in the sample. This compound will not interfere if used for non-chlorinated water.

When collecting water samples, first try to get "historical" samples that might give an indication of the condition of the water at the time it was ingested by those who became ill. Obtain historical samples from water in bottles in refrigerators, toilet tanks, hot water tanks (for chemical analyses only), fire truck reservoirs, storage tanks, and taps at seldom-used and dead-end locations, and from ice in refrigerators and commercial ice plants. Direct the laboratory to test historical samples for pathogenic organisms or toxic chemicals, as well as indicator organisms, because these samples have a chance of still containing the etiologic agent, whereas samples collected during the investigation several days or weeks after the event may be of water that has been flushed free of contamination or has been significantly diluted.

Take samples from 8 to 10 points throughout the distribution system. Sample dead-end locations if they are found. Do not neglect to obtain raw water samples even though treatment is provided. This is important, as it suggests possible sources of contamination and reflects the effectiveness of treatment. Compare these test results with records of results on previous samples of raw or treated water.

Before drawing a sample from a water tap, make sure the tap is connected to the supply to be tested. Do not collect samples (other than for *Legionella*) from hose connections, sprays, or swivel faucets; uncouple these connections or choose different outlets. It is unnecessary to flame outlets, as this does not improve the quality of the sample. First, ensure your hands have been thoroughly washed then take a line sample by allowing the water to run to waste for 5–10 min. Adjust the flow of water so that the thiosulfate will not wash out of the bottle or bag (do not overfill — most laboratory bottles indicate a maximum fill line). Keep sample containers closed until the moment they are to be filled. Hold the bottle near the base, fill to the "fill line" or within an inch of the top without rinsing, and immediately replace the stopper or cap and secure the hood, if attached. If a Whirl-pak™-type plastic bag is used instead of a bottle, hold the base, rip off the perforated top, open the bag by pulling the side tabs apart, grasp the end wires, and place the bag under the flowing water. Remove the bag before it is completely filled and squeeze most of the air out; fold over the top of the bag several times and secure by twisting the end wires. Take a source or a distribution line sample by opening the tap fully and letting the water run to waste for sufficient time to empty the service line (or if in doubt, for 5 min) and proceed as above.

Collect samples from open shallow wells and step wells by dropping a clean wide mouth container on a string or rope into the well. Allow the container to sink below the water surface and then pull it out of the well. Pour contents into a sample jar or bag.

Collect samples from rivers, streams, lakes, reservoirs, springs, toilet tanks, and non-pressurized storage tanks by holding a 200 mL sample bottle near the bottom and plunging it neck down to a depth of 15 cm (6 in) below the surface; turn it right side up, and allow it to fill. Don a plastic disposable glove when small vessels used for drinking are sampled in this manner. When collecting these samples, move the bottle in a sweeping, continuous, arc-shaped motion, counter to stream flow or in a direction away from the hand. Collect samples at locations approximately one-quarter, one-half, and three-quarters the width of the stream or water course. Special apparatus can be used for sampling at various depths. Samples can then be taken by positioning large bottles on a rod or pole at the desired depth and location before pulling their stoppers with a wire, string or thin rod. Samples of bottom sediments are sometimes useful for the detection of certain pathogens. Collect surface scum or regions containing dense particulate colored material when seeking cyanobacteria (blue-green algae). Collect slime, if present, when seeking *Pseudomonas*. If large amounts of water are needed, seek assistance and obtain specialized sampling equipment from agencies responsible for water quality.

If possible, avoid wading when sampling bodies of water because wading often stirs up bottom sediments. If this is the only way to get a sample, however, wade against any current (e.g., upstream in creek or river) and keep moving forward until sample taking is completed. Piers or similar structures, or the front end of a drifting or slow moving boat, make good sampling stations.

Concentration of bacteria by the use of swabs, filters, or by absorption, is particularly important when waterborne pathogens are sought. To concentrate bacterial pathogens from flowing water (e.g., streams, lakes, sewer lines, or drains), suspend Moore swabs (or non-medicated sanitary napkins or non-medicated tampons if Moore swabs are unavailable) for 3–5 days. These can be held in place by wire just below the surface or at other depths. If rodents are about, put Moore swabs in wire baskets. After the sampling period, either put swabs or pads into a plastic bag and pack in ice, or put the swabs or pads directly into an enrichment broth for the pathogen sought. Take or send these to the laboratory promptly.

Concentration of microorganisms can be increased by filtration with a variety of filters (e.g., membrane filters, cartridge filters, or other filter media). When membrane filters are used for pathogenic bacteria recovery, pass at least 1 L of water (relatively free of turbidity) through a sterile 0.45 μm membrane filter. For viral analysis, use virus-absorbing electropositive cartridge filter to concentrate 400 L or more water (see Table F). Keep filters cool (but not frozen) and ship to a reference laboratory for further processing. For *Giardia* cysts and *Cryptosporidium* detection, collect samples by passing at least 400 L water through a cartridge filter (see Table F).

For inorganic chemical analyses, use 1 L polyethylene containers. These should be new, or acid-washed if previously used. Collect the water without flushing the lines, preferably in the early morning before water is used. For trace metal analyses, preserve one sample with 2 mL of high-grade nitric acid to a pH of 1 or less. This is particularly important whenever it is suspected that metals may have leached from water pipes or vessels. For organic chemical analysis, use 4 L glass containers with teflon-lined caps. Clean and rinse the containers with a good quality laboratory solvent and heat at 400°C for 20 min. Rinse the cap thoroughly with distilled water. Fill the container so that there is a minimum of air space. For physical analyses, collect at least 2 L, or other amounts requested by the laboratory.

Collect ice aseptically in sterile plastic bags or jars. Use sterile tongs to collect cubes; sterile spoons for collecting chipped or crushed ice; and sterile chisel, hammer, or pick to chip block ice. Put block ice or large chips into plastic bags.

If *Legionella* is sought, sample water at sites of any source that may have been aerosolized and send to a lab with proven expertise in *Legionella* isolation and characterization, such those in the CDC ELITE Program. This includes cooling towers, evaporative condensers, water heaters and holding tanks, humidifiers, nebulizers, decorative fountains and whirlpool baths (see section on investigating sites where aerosols are disseminated for a more complete listing). Turn off fans of condensers before sampling; if this is not possible, wear a respirator. Use 250 mL to 1 L polyethylene bottles that have had sodium thiosulfate added if the water to be tested has been chlorinated. For each sample, don disposable plastic gloves and collect the sample by inverting the bottle and moving it in a continuous arc away from the hand. Measure and record water temperature. Handle samples as described in Table F. Rub swab over faucet aerators and shower heads if these are considered as sources of aerosols. Break stick and allow tip to fall in a tube containing 3–5 mL sterile water (not saline).

Investigators are often requested to test air to demonstrate the presence of *Legionella* in aerosols. Although legionellosis is an airborne disease, legionellae are susceptible to low humidity and become non-viable on drying. Therefore, air sampling is an ineffective and inefficient way of determining whether a *Legionella* hazard exists, and it can thus be misleading.

Label each container with sample number, date, time of collection, and your name or initials. Complete the Water/Ice Sample Collection Report, Form F, for the first sample. List additional samples with sample numbers and other pertinent information on the back of the form. In those situations where the laboratory needs additional information, attach the appropriate G series forms. Send the original Form F and list with samples to the laboratory; retain a copy for your files. Inform the laboratory of the type and number of samples and specimens; also, consult with the laboratory on methods to preserve and transport samples, if necessary, and on time of their arrival.

If legal proceedings are anticipated, deliver sample personally to the analyst, or seal the sample container in such a way that it cannot be opened without breaking the seal. Note on Form F the method by which the bottle was sealed. Maintain a chain-of-custody log to document the handling of the sample, and have the log signed and dated each time it changes hands. Consult with state/provincial regulatory agency on complying with legal requirements for chain-of-custody procedures. Recipient should record on the form whether the sample was sealed when the laboratory received it.

If analysis cannot be done on the day of collection, chill water samples rapidly and hold them at temperatures at or below 4°C (39°F), but Do NOT FREEZE, because populations of bacteria such as *Escherichia coli* and of parasites decrease during frozen storage. Hold ice samples frozen; if this is not possible, keep the temperature below 4°C h (39°F).

How to Transport

Investigators should consult with the testing laboratory that will be used to receive specific laboratory sample packaging, labeling, and transportation instructions as protocols are dependent on specific transportation regulations (IAFTA, TDGR) within each jurisdiction.

Ensure each sample is uniquely identified and labeled (as per the receiving laboratories requirements). Many laboratories include barcode labels along with the sample containers within the sample collection kits. Ensure that the correct label is affixed onto the correct sample container and that this information is transferred to the shipping manifest accurately (chain of custody form). Specimens should be packed and the packages labeled according to applicable regulations governing transport of hazardous materials.

Generally, the transport of samples of water and ice intended for laboratory analyses are packed and shipped in a manner to ensure the sample does not change from the time of sampling to the time received by the testing laboratory and shipped using the most expeditious means (e.g., personal delivery or overnight mail). Typically samples of water or ice are packed with refrigerant (ice packs, dry ice, etc.) in insulated and sealed containers (see Table F).

Receipt of laboratory analysis. Record results of laboratory test samples on Form I.

How to Take On-Site Measurements

Several measurements are routinely called for during on-site investigations. Brief instructions are given for those that are commonly done; nevertheless, follow manufacturer's instructions if these are available.

Measure free, combined and total residual chlorine and other disinfectants. Color comparison kits are available for testing for free, combined and total residual chlorine. The diethyl-*p*-phenylenediamine (DPD) test is an example (see Table A). Check instrument calibration regularly. Use dry reagents, because the liquid forms are unstable. Chlorine comparators can be used to test for bromine by multiplying the result by the factor 2.25 and to test for iodine by multiplying the result by the factor 3.6.

Measure temperature. Measure water temperature by immersing the sensing end of either thermocouples, transistors, or thermometers into the water. Sometimes measurements need to be made at various depths; use thermocouples with wire leads of sufficient length for this purpose. Calibrate temperature measuring devices periodically.

Measure pH. Calibrate the pH meter as recommended by the manufacturer with at least two standard buffers (e.g., pH 7.0 or 10.0) and compensate for temperature, if the meter does not do it automatically, before each series of tests. Remove a sample of water to be tested and immerse the pH electrode into the sample; record the reading. pH can also be measured by color comparators that employ color indicator solutions or discs. (Ranges of pH color indicator solutions are bromophenol blue, 3.0–4.6; bromocresol green, 4.0–5.6; methyl red, 4.4–6.0; bromocresol purple, 5.0–6.6; bromothymol blue, 6.0–7.6; phenol red, 6.8–8.4; cresol red, 7.2–8.8; thymol blue, 8.0–9.6; and phenolphthalein, 8.6–10.2.) In this case, water containing more than 1 mg/L chlorine in any form must be dechlorinated with sodium thiosulfate before the pH indicator solution is added to prevent decolorization of the indicator. Always report temperature at which the pH is measured.

Measure turbidity. Nephelometric Turbidity Unit (NTU) is the usual standard unit, but other turbidity measurements (such as particle counts) are used. The NTU requires a nephelometer, which measures the amount of light scattered predominantly at right angles and absorbed by suspended particles (e.g., clay, silt, finely divided organic matter, inorganic matter, soluble colored organic compounds, and microscopic organisms) in the water sample. Calibrate turbidimeters with a standard reference suspension. Make turbidity measures on the day samples are taken. Vigorously shake samples, wait until all air bubbles have disappeared, and then pour sample into turbidimeter tube. Read directly from scale on instrument or from an appropriate calibration scale.

Measure air flow. Pump chemical smoke into the air at the exit of the device suspected of releasing aerosols. Observe the direction and spread of the smoke. Otherwise, measure pressure differentials with a micromanometer.

Measure other attributes of water. Follow instructions given by manufacturers or in standard reference books (see Further Reading).

Trace and Confirm Source of Contamination

Use fluorescein dye, lithium or other tracers in appropriate soils to determine the means by which contamination from sewage, industrial wastes, or other sites of pollution reached the water supply. Fluorescein dye is particularly helpful in evaluating flow of contamination through fissured rock, limestone, gravel, and certain other soils. This dye is not readily absorbed or discolored by passage through these soils or sand, as are many other dyes, but it is discolored by peaty formations or highly acid (pH < 5.5) soils.

Make a concentrated fluorescein dye solution by mixing 300 g of fluorescein powder into a liter of water. Usually, 2/3 to 3 L of this solution are sufficient for the test for up to 60,000 L of water. Fluorescein dye is also available in liquid and tablet form. One tablet will dye approximately 480 L (~120 US gal).

Pour the calculated amount of fluorescein solution or put a sufficient number of fluorescein tablets into a receptacle at a point of potential pollution. Usually this point will be located within 100 yards and at a higher elevation than the water source under study. Cesspools, latrines, distribution boxes, sink holes, borings, septic tanks, drains, manholes, toilets and plumbing fixtures are typical places to introduce the dye. If dye is poured into a plumbing fixture or dry hole or boring, add water to wash it down. The amount of dye to use varies with the distance the dye must travel, the expected time of the journey, the size of the aquifer or water channel, and the nature of the soil.

Take samples of the water when the dye is introduced into the test hole or fixture and then hourly for up to 12 h to detect arrival and departure of fluorescein. If no dye is observed, repeat the test with twice the amount of dye. Whenever possible, use a fluorescent light or fluorometer to analyze water samples for evidence of fluorescein. A fluorometer can be set up and calibrated, and a continuous recording can be made. This meter can detect fluorescein in concentrations of μg/L (ppb). Fluorescein dye will temporarily color water, which discourages use of the water until the dye is sufficiently degraded or diluted. Alternate tracers can be used if specific ion meters are available.

The dye stains all it touches. Methanol is a good solvent for the dye, and hypochlorite solutions decolorize it; both can aid in removing stains. Abrasive soaps are useful for cleaning stained skin; fluorescein-stained clothing should be washed separately.

Appearance of dye in a water supply is conclusive evidence of seepage from the site where the dye was introduced. Failure to detect dye, however, is not conclusive evidence that seepage did not or would not occur if more dye had been added or if weather conditions or subsurface flow had been different at the time of the test than during the outbreak event.

Illustrate source and direction of contaminated water flow as indicated by the dye test on Form G1. Take photographs of sources of contamination and evidence of staining of the ground at the site or dye-stained color of the water. In situations where a single source of contamination is obvious or where multiple sources are readily apparent, dye studies serve little purpose.

Water not Intended for Drinking as a Source of Illness

Drinking water, however, is not the only source of water that may contribute to outbreaks. Other sources of water that can contribute to outbreaks include water not intended for drinking, recreational water and water used in agriculture during harvesting and packaging.

Legionnaires disease is the pneumonia caused by the inhalation of contaminated water aerosol containing the bacteria *Legionella*, with *Legionella pneumophila* being responsible for 85% of all infections. It is also a common cause of healthcare associated pneumonia. *Legionella* can replicate within free-living amebae in water, allowing it to resist low levels of chlorine used in water distribution systems. Risk of infection is more common in warm and humid weather, when water droplets are able to drift further due to higher absolute humidity.

Fifty percent of all *Legionella* outbreaks have been traced to cooling towers with *L. pneumophila* serogroup 1 responsible for all cooling tower outbreaks. All aerosol generating devices, however, can be potential sources of *Legionella*. Some other sources of aerosolization that may have contributed to or be associated with outbreaks include: whirlpool displays, building's air conditioning systems, water spray fountains, public bath houses, vegetable misting systems in grocery stores, evaporative condensers, showerheads, humidifiers, air scrubbers, car washes, ornamental and decorative fountains, potting soil, respiratory therapy equipment, dental units, road asphalt paving machines, car windshield washer fluid and car air-conditioning systems.

In the investigation of a *Legionella* outbreak, (See Box 3, The Flint Water Crisis, which describes a likely *Legionella* outbreak from a commercial water source) due to the varied sources, there is a need to use a broad investigative questionnaire and the collection of environmental data. Environmental factors such as dry bulb temperature, relative humidity and wind rose data can provide information regarding drift evaporation, deposition (settling) and the size of the affected zone. Although aerosol drift can carry *Legionella* up to 6 mi (10 km), the risk of infection is usually highest within 1600 ft (500 m) of the source. There are also air dispersion models that can be used to determine drift zone and the use of Human Activity Mapping in the identification of potential sources.

Recreational Water

In general, *E. coli* and norovirus are the most common pathogens responsible for recreational waterborne outbreaks associated with non-treated water such as beaches and lakes. *Cryptosporidium*, which is resistant to chlorination, is the most common pathogen resulting in outbreaks in treated water venues such as swimming pools and water spray parks. It should be noted that *E. coli*, the indicator of choice of recreational water samples, is not indicative for the presence of norovirus and *Giardia*, *Cryptosporidium*. *E. coli* can also be "naturalized" and have been found to survive and multiply in beach sand. Beach water sampling results therefore may provide false positive or false negative results and may not be the best indicator for the presence or absence of pathogens.

Recreational waterborne outbreaks are not just traced to the ingestion of contaminated water (Table C. Illnesses acquired by contact with water: A condensed classification by, symptoms, incubation period, and types of agents). Hot Tub Rash, or *Pseudomonas* Dermatitis/Folliculitis commonly occurs in public hot tubs or spas such as those found in hotels. The rash is often a result of skin infection from the bacteria *Pseudomonas aeruginosa* colonizing in the hair follicles after exposure to contaminated water. *Pseudomonas aeruginosa* is an opportunistic pathogen that can survive within the biofilm on the tub surface or within the piping system. Outbreaks can occur when there is a heavy bather load resulting in an increase in chlorine demand, which in turn reduces the effectiveness of the disinfectant to control the population of *Pseudomonas*.

Blue-green algae or cyanobacteria bloom can occur in warm, slow-moving or still water. When conditions are favorable, mostly during hot summer weather, cyanobacteria populations may increase dramatically, resulting in a "bloom" as they rise to the surfaces of lakes and ponds. They resemble thick pea soup and are often blue-green in color. Although blooms can occur naturally, water bodies which have been enriched with plant nutrients from municipal, industrial, and agricultural sources are particularly susceptible. Some cyanobacterial species may contain various toxins, some are known to attack the liver (hepatotoxins) or the nervous system (neurotoxins); others simply irritate the skin. Health effects from cyanotoxin exposure may include dermatologic, gastrointestinal, respiratory and neurologic signs and symptoms (Table B. Illness acquired by ingestion of contaminated water: A condensed classification by symptoms, incubation periods, and types of agents).

Irrigation and Processing Water

Water can also be an indirect cause of foodborne outbreaks by providing a media for the survival, transportation and the introduction of pathogens into food products. Water used during production, including irrigation, pesticides and fertilizers application and washing, frost protection, harvesting, has long been recognized by food safety scientists as one of plausible and probable sources of the contamination of fresh fruits and vegetables. There have been many outbreaks from produce traced to pathogens being introduced by contaminated irrigation water. Although harvested products are sometimes washed with chlorine solution, pathogens may still survive the process through internalization. *E. coli* O157:H7 may migrate to internal locations in plant tissue and be protected from the action of sanitizing agents by virtue of its inaccessibility. Experiments have also demonstrated that *E. coli* O157:H7 can enter the lettuce plant through the root system and migrate throughout the edible portion of the plant. However, this claim has been refuted by others. *Salmonella* and *E. coli* can also adhere to the surface of plants, and enter through stomata, stem and bud scars and breaks in the plant surface caused by harvesting and processing. Water containing bacteria can be drawn into the produce if it is immersed in or sprayed with water that is colder than the produce itself. *E. coli* O157:H7 may also use its flagella to penetrate the plant cell walls and attached to the inside of the plant. Once attached, it may be able to grow and colonize the surface of the plant. The concerns are not just with bacteria. The present of norovirus in the hydroponic water can result in internalization via roots and dissemination to the shoots and leaves of the hydroponically grown lettuce.

Irrigation water may be contaminated from runoff from nearby domesticated animals and their lagoons, feedlots, ranches into rivers; from feral/domestic animals with direct access to creeks, ditches, rivers, ponds; from sewage flows into waterways and contaminated wells. In some parts of the world sewage contaminated water is preferred for irrigation despite a potential risk of transporting enteric pathogens, since it carries nutrients (N and P) for the plants.

There is sufficient information to conclude that the application method of irrigation water to fresh produce can have an effect on the microbiological risks associated with the crop. In general, keeping water away from the edible parts of ready-to-eat crops that are consumed without cooking can result in a lowered risk of a foodborne outbreak.

The least to more risky methods for irrigations for microbial contamination are:

Subsurface irrigation (buried soak hoses) < drip irrigation < indoor flood irrigation (hydroponics) < outdoors flood irrigation (water-filled furrows) < overhead irrigation (sprinklers).

Collection and Analysis of Data

The Epidemiological Approach

An outbreak of illness arising from exposure to water demands immediate epidemiological investigation to assess the situation, gather, evaluate, and analyze all relevant information, with the goal of (1) halting further spread of this illness, and (2) predicting, preventing, and/or attenuating future outbreaks. This twofold mandate of epidemiology is usually described as "*surveillance and containment.*"

At the commencement of an investigation, the *unknowns* usually outnumber the *known* facts. There is no substitute for prompt, thorough, and careful collection of interview data from ill and well persons who ingested or contacted the suspect water, attended a common event, or who were part of a group of persons where illness occurred. Careful analysis of these data, particularly with reference to common patterns of "time," "place," and the characteristics of the persons involved, can often eliminate many vehicles, agents, and pathways quite early in the investigation, and focus on the remaining possible vehicles, routes, and agents. Later, laboratory results may confirm the agent, the specific pathology, the route taken by the infection or toxic agent, and indicate what is needed to stop the spread, but early epidemiology can often be invaluable in predicting the outcome and taking preventive steps to contain the problem before the lab results are available. Lessons can be learned from most outbreak investigations and are invaluable for increasing our understanding of these pathologies, and preventing their future occurrence.

Determining an Outbreak

An *outbreak* is defined as either *an unusually large occurrence of an expected illness* at that time of year in that place, or *the occurrence of a type of illness that does not usually appear* at that season and location. The "time" factor should be studied immediately by plotting the onset time of each case on a time-based grid, to create the *epidemic curve*. Although any number of cases can be involved, the minimum number for an *"outbreak"* to be declared is *two associated cases*, with special exceptions such as *Naegleria fowleri* where, because of the severity and the possibility that cases may have been missed, *a single case* constitutes an "outbreak." Although the epidemic curve is usually measured in hours or days, protracted exposure to agents in water may mean apparent sporadic cases linked to a common source over months or years.

The "Case Definition" and Its Importance for the Analysis

If an "outbreak" is suspected by a sudden increase of cases, determining who is to be categorized as a "case" is not necessarily a simple process. Many people notoriously fail to report enteric illness for many reasons: embarrassment, lack of time, no clear idea which agency should be notified, mild self-treatable symptoms, or simply because they prefer not to make a fuss. They may therefore be incorrectly classified at least initially as *"non-ill."* Consider also that 4–6% of the general population will have experienced some form of "upset stomach" in the last 24 hours, regardless of exposure to the suspect item, and they may be incorrectly classified, at least initially, as *"ill."* To reduce the *"false negatives"* and *"false positives"* that are expected with self-reporting, the investigator needs to establish a working case-definition.

A careful case definition categorizes people as *"case"* or *"control"* with the best accuracy possible within the time constraints and resources available. A case definition could be considered *"too sensitive"* if it classifies as a *"case"* a person who experiences: *"… at least one episode of stomach cramps, nausea, vomiting, or diarrhea in the last 48 hours."* This would confuse subsequent analysis, and produce more false positives. Similarly, a case definition could be considered *"too specific"* if it classifies as *"not-ill"* a person who had experienced only three episodes of diarrhea or vomiting, because they failed to satisfy a case definition requiring *"…at least four episodes of vomiting or diarrhea in the last 48 hours."* Should this last individual, having been declared as not fitting the case definition, be taken into the *"not-ill"* group, the error and subsequent analysis is confounded even further.

A reasonable case definition therefore attempts to reduce both types of errors, and will depend upon the early indications of what the etiology may be. In the instance of a suspected salmonellosis, a case could be defined as *"A person who was in good health before attending the event on Monday May 3rd, and who experienced two or more of the following symptoms anytime up to midnight, Sunday May 9th.: nausea, vomiting, stomach-cramps, diarrhea, headache, or fever."* Note that a case definition should include a *place* of exposure if known, a *timeframe* during which symptoms may have been experienced (salmonellosis has a range from 6 to 72 h. usually 24–30 h), and the additional footnote that the individual was not already symptomatic before the suspected "exposure."

Table 4 Symptom profile

	Number of cases	Percentage reporting each symptom
Diarrhea	195	95
Abdominal cramps	182	89
Nausea	52	25
Vomiting	42	20
Fever	6	3
Headache	2	1
Total cases	205	

The Symptom Profile

Calculate the percentage of ill persons who manifest each symptom by dividing the number of persons reporting the given symptom by the number of cases (205 for the example, Table 4) and multiplying the quotient by 100. The distribution of symptoms can be used to identify the most likely pathogen, and aids in requests to the laboratory for microbiological assays of samples and specimens. Other symptoms (e.g., prostration, lethargy, weakness) may be included if deemed appropriate or helpful, but the six symptoms in Table 4 should always be included. Headache, for instance, is associated with many viral infections (e.g., norovirus, rotavirus), but much less so with bacterial infections. Fever is usually associated with an invasive bacterial infection (such as salmonellosis or campylobacteriosis), and is not usually seen in outbreaks of simple enteritis (such as with cholera).

This information helps to determine whether the outbreak was caused by an agent that produced intoxication, an enteric infection, or generalized illness. In the example given, a predominantly diarrheal syndrome without much fever or headache tends to eliminate some of the viral infections (norovirus or rotavirus) or the host-adaptive/invasive serotypes of *Salmonella* (e.g., *S.* Dublin or *S.* Choleraesuis). Median onset time calculations may further reduce possible candidate etiologies. In historical investigations, or where no laboratory confirmation is possible, the symptom profile and onset times can sometimes predict the etiology of the outbreak within reasonable certainty.

The Epidemic Curve

Plotting the Cases

An *epidemic curve* (also called an *onset curve* or *onset distribution*) is a graphic illustration called a histogram that shows the distribution of the time of onset of first symptoms for all cases that are associated with the disease outbreak. Paper printed with square "grid" lines will allow the investigator while on site to represent each

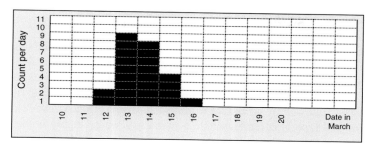

Fig. 2A Onset histogram for 24 cases of acute enteritis from March 12 to March 16, illustrating a point-source without propagation

case as a single "block." The horizontal axis is the sequence of intervals of time and date. The unit of time that defines the width of each interval depends on the characteristics of the illness under investigation. For example, intervals of days or weeks are appropriate for diseases with long incubation periods, such as cryptosporidiosis or hepatitis A. Intervals of a day or half-day are appropriate for outbreaks of enterohemorrhagic *E. coli* strains or shigellosis, while single-hour, 4-h, or 6-h intervals will be more suitable for illnesses with shorter incubation periods, such as chemical poisonings. The vertical axis is always the actual count, or "*frequency*" of cases (blocks) stacked at each interval. It is often necessary to redraw the onset curve as more accurate information becomes available.

If the illness is known, a rule of thumb is that the time interval used for each "block" on the *x*-axis should be no more than ¼ the incubation period of the disease under investigation. If the illness is not known, select an interval where the data produces a bell-shaped curve; not too flat and not too tall. Construct this graph using time-of-onset data from Forms C or D, employing an appropriate time scale.

Means, Medians, and Modes

Once all the onset times for the cases have been plotted on the histogram, determine the *range* as the interval between the shortest and longest incubation periods. In Fig. 2A, the range is the 5 day period from the 12th to 16th March. The *median* onset time is preferred to the *mean* because the latter is vulnerable to a few or even a single very small or very large value. The *median* on the other hand, is the *mid-value of a list of all individual onset times*, including duplicate entries, that are ordered in a series, from shortest to longest. If the series comprises an even number of values, the median is the mean of the two middle values. Most standard reference texts on communicable diseases give onset times as median values.

The *mode* is simply the <u>*interval*</u> *having the largest number of observations.* A distribution with a single "peak" is called a *uni-modal distribution*, while an outbreak with two peaks is called "*bi-modal*." Subsequent modal peaks following the first may indicate either a "secondary wave" of cases or the exposure of *other* people at a *later* time.

Interpreting the Epidemic Curve

The shape of the epidemic curve helps to determine whether the initial cases originated from a single ***point-source*** exposure (such as water or food available for only part of a day), or from repeated exposures for a longer time, or even more gradual person-to-person spread. A *point-source epidemic curve* is characterized by a sharp rise to a peak, followed by a fall that is almost as steep (Fig. 2A). An "explosive" outbreak of this type is common where a municipal water supply is the vehicle, affecting large numbers of people in a very short period of time, but without secondary cases occurring, or any evidence of onward spread within the community.

Propagated outbreaks are those in which the initial victims (*"primary cases"*) manage to spread the agent to other people (*"secondary cases"*) such as family members, patients, clients, or other contacts in crowded places through aerosols, personal contact, or contaminated water/food/utensils/surfaces, etc. Propagation following a point-source exposure is demonstrated by a second increase in reported cases following the decline of the first cluster. Sometimes this takes the form of a second "modal peak" separated by approximately one incubation period, but this distinction is soon lost. Figure 2A shows no evidence of propagation; Fig. 2B suggests that propagation *may* have taken place, although care must be taken to consider other explanations.

In addition to (1) true propagation, where the secondary wave can be expected to appear one incubation period after the first, secondary waves may be also explained by (2) exposure to the *same point source* (e.g., food or water supply) at *different, but specific times* by other people; this might be a repeated offering of contaminated food or water at two or more mealtimes; (3) a second pathogen (perhaps from the same unhygienic food or water source) which may have a different symptom profile and a *different (incubation) time.*

Slow propagation from the *beginning of an outbreak* with neither an obvious point-source, nor any distinctive "waves" separated by an incubation period as in Fig. 2C, usually indicates one-at-a-time person-to-person spread through close-contact, poor personal hygiene, aerosol (e.g., influenza, or SARS), or sexual transmission (e.g., HIV/AIDS). It can also be explained by (non-propagated) continuing

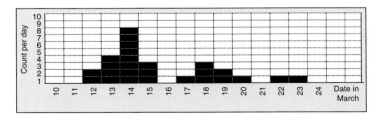

Fig. 2B Onset histogram for outbreak of enterohemorrhagic *E. coli* (EHEC) enteritis, March 12 to March 23. Shown are 16 primary cases, 7 secondary cases, and 2 tertiary cases. Point-source with propagation

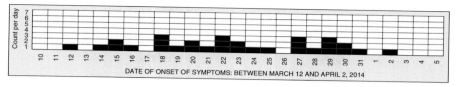

Fig. 2C Onset histogram for 30 cases of shigellosis, March 12 to April 2, illustrating slow spread through a community through either propagation (person-to-person spread via poor hygiene), or exposure by many people to a small well at different times (a non-propagated route)

exposure, for example drinking of contaminated surface water following a conflict, natural disaster or other breakdown of infrastructure. As such it is commonly associated with waterborne cholera, shigellosis, typhoid fever, or *E. coli* infection, and characterized by scattered cases which continue until the chain of infection is cut. Slow, constant and/or intermittent exposure to persons over time to pathogenic microorganisms can also result from sewage run-off after a series of heavy rainfalls.

Estimating the Incubation Period Where the Exposure Point Is not Known

In addition to revealing whether the outbreak was due to a single point-source, or had been spread steadily through the community by propagation in some way, another important objective in constructing the epidemic curve is to estimate the incubation period of the illness if it is not already known. With waterborne illness especially, the time of exposure may be further obscured because people usually drink water several times a day. Hence, the incubation period cannot always be determined for each case, but the actual time of onset is usually available.

The incubation period is the interval between exposure to food or water that is contaminated (with enough pathogens or with a sufficient concentration of toxic substances to cause illness), and the appearance of the first sign or symptom of the illness. Each etiology is characterized by a typical incubation period (Tables B, C, D, and G). Individual onset times will vary due to immune factors, co-morbidities, the dose ingested, and other ingested materials, but the investigator can often make a rough estimate of the average incubation time by examining the aggregation of all onset times as an epidemic curve.

The *modal peak* of a single "cluster" or distribution is the time interval in which most cases commence symptoms. In Fig. 2A this occurs on March 13, and in Fig. 2D that occurs at the double interval Feb 10–11th. Where two *separate* modal peaks (a "*bi-modal distribution*") suggests secondary cases ("propagation"), then the distance between the first two modes is a good estimate of the incubation period. Figure 2B shows about 4 days between primary and secondary modal peaks, suggesting that the initial exposure is likely to have been 4 days before the first mode. In Fig. 2B this would be sometime on or near the 10th of the month.

If the exposure point is *known* but the agent is not, then that estimation of the *median incubation period* will allow many etiologic agents to be excluded due to

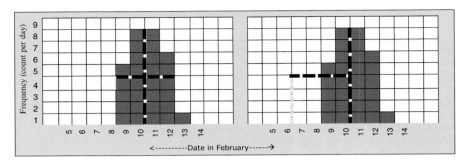

Fig. 2D Cases ($n=28$) commencing during 1-day intervals in February, 2013. If the curve rises rapidly to a peak and drops sharply, (1) draw a vertical line from the modal peak to the x axis; (2) mark the point half-way up that height; (3) draw a horizontal line to show the width of the peak at that point; (4) slide the horizontal line to the left such that the right end touches the vertical line exactly. (5) A perpendicular dropped from the left of the horizontal line will be the best estimate of the exposure time on the horizontal axis. In this case it was the evening of February 6th

incubation periods that are clearly outside the range of times observed. The list of possible candidates can be further reduced by examining the symptom profile and other characteristics of the illness and suspect food or water vehicle. As time passes, the onset curve also provides an ongoing measure of the potential for propagation, and the incidence rate. All this information can be useful in deciding whether the illness in question is an infection or intoxication and thereby determining which laboratory tests should be requested (Tables B, C, D, and G). Note that not all water or foodborne illnesses listed in a standard reference such as the *"Control of Communicable Diseases Manual"* (APHA 2014), are *directly* communicable *person-to-person*; many require a suitable substrate (food or drink) and adequate time/temperature combinations to attain sufficient numbers or the production of enough toxin to induce a pathological condition.

An exposure time can sometimes be estimated from a clear, point-source, single-exposure onset distribution (Fig. 2D). It has no solid basis in statistics, but has sometimes been found to be useful in practice.

The typical incubation periods for most foodborne and waterborne illnesses are readily available for comparison (e.g., *Control of Communicable Diseases Manual*, APHA/CDC, 2014), and in this manual in Tables B, C, D, and G.

Calculate Incidence, Attack, Exposure Rates for Groups Affected

Overall Attack (Incidence) Rates

An incidence rate is the number of *new* cases of a specified disease reported during a given time period in relation to the size of the population being studied, multiplied by a constant, usually 100, to give percentages. Thus 14 new cases of *E. coli* O157:H7 infection among the 140 residents of a children's summer camp in July is an incidence rate of $(14/140) \times 100$ or $(0.10) \times 100 = 10.0\%$ for that month. If several people have left and their state of wellness is *not known*, their impact should be expressed in the form of the *possible range of values* around the known incidence rate within the two extremes whereby they may *all* be well or they may *all* be ill. Thus, where six children who were at the camp had departed around the time of the outbreak and their health is unknown, the range could be from a possible $(14/146) \times 100$ (or 9.6%) if all of the six had been well, up to $(20/146) \times 100$ (or 13.7%) if all had been ill. Note that the "missing" six are added to the denominator only when we speculate that none were ill, whereas they are added to the numerator AND denominator if we speculate that they might all have been ill. In this example, the *overall incidence rate* would be reported as "*10%, with a possible range from 9.6 to 13.7%.*"

Factor-Specific Attack (Incidence) Rates Where Possible

Depending upon the situation, it is often necessary to identify exposures which may be related to the illness, and to calculate an incidence rate for each such exposure. For example, in the summer camp illustration (above) it might be useful to enquire if gender, age, location, or some other attribute or activity increased the risk of becoming ill. This should not be interpreted automatically as implying that a given exposure would be associated with the outcome in any situation. By hypothesizing that gender was linked to the risk of illness, for example, does not imply that males are more vulnerable to the illness (the outcome) than females, but it can indicate that gender may have been *related to the exposure*, which in turn increased the risk. As an illustration, suppose that boys at the camp had been swimming, while the girls had gone on a nature walk. The boys may subsequently show increased incidence rate for *E. coli* O157 infection, not because they are more susceptible, but because of their activity. Every proportion or percentage statement should be made with clear reference to the appropriate denominator used.

Incidence rates of waterborne illnesses are usually similar for both sexes at any given age group in the population, but differences in activities or dietary habits or susceptibility due to age or underlying health status can change the risk. The very young, the elderly and the immunocompromised can be at more susceptible, while in some instances, previously exposed populations may have developed a measure of immunity to an infection that may still cause more serious illness among visitors.

A further complication arises where the "at-risk" population (perhaps residents at an institution, summer camp, or on a cruise ship) have generally consumed *all* the food and water for the extended period. Careful interviewing of affected persons often uncovers one or more persons who entered the subject community shortly before becoming ill or who visited the community for a short time and became ill after leaving it.

Attack (Incidence) Rates by Place of Residence

Example: The south-west part of the county is served by three semi-private water systems. Thirty cases of waterborne illness are being investigated in the area. When the numbers of cases are displayed for each water system, no clear grouping or clustering is evident, although the **Delta** supply appears associated with about 50% more cases than the other two (Table 5A).

However, when the analysis introduces the total population of persons who depend upon each water system (as denominators), a different scenario emerges. The incidence rates (expressed here as percentages) now allow a meaningful comparison (Table 5B). We can see that persons using the **Bravo** system have roughly *five times the risk of illness* compared to people who are served by the other two systems. The use of the denominator is vital for most calculations. Caution: Numerous other factors may also explain the outbreak and these should be carefully examined. For example, the households using the **Bravo** supply may be closer to an unhygienic corner store, drink from a cross-connected public water fountain, or their children may swim in a more polluted pond than the other communities. Potential sources such as these should be eliminated before the water supply is announced as the source of the illness.

Sometimes a *spot map* may be useful in showing the location of the residence of each case, while on a larger scale, the rates of illness can be shown using city blocks, census tracts, townships, or other subdivisions. Different colors or symbols to indicate cases with different time of onset periods (such as weeks) may help to support a hypothesis as to where contamination was introduced, inasmuch as the earliest cases tend to cluster around the point where contamination first occurred. The weakness of this procedure is that if the exposure had been at a restaurant, workplace, or school, plotting the relationship to the location of the home would not be useful.

Preparing to Calculate Associations Between Exposure(s) and Illness

The investigation of waterborne or foodborne disease outbreaks invariably commences *after* both exposure and illness have happened. This is the classical "***case-control***" study, where a group of ill people ("*cases*") and a group of non-ill people

Table 5A Comparison by numbers of cases (no denominator)

System	Alpha	Bravo	Delta	Total
Cases	9	8	13	30

Table 5B Comparison by rates (using denominator)

System	Alpha	Bravo	Delta	Total
Cases	9	8	13	30
Population supplied	360	64	496	920
Attack rate	2.50%	12.5%	2.62%	3.26%

Table 6 Exposure and outcome data arranged in 2×2 table

	Ill (cases)	Not ill (controls)	Total
Exposed to X	25	8	33
Not exposed to X	12	22	34
Total	37	30	67

("*controls*") are compared in terms of their exposures.[1] To measure the association between exposure and illness, the data are typically displayed in a 2×2 contingency table. Table 6 compares 37 cases and 30 controls in terms of their *exposure* to a suspected factor "*X*." The table is ready for analysis using odds ratio, as well as the chi-square or the Fisher's exact tests where appropriate. One 2×2 table will be used for each possible exposure (e.g., each beverage, food, or other material).

As many cases as can be identified and contacted, and as many non-ill people (controls) as can be found, should be interviewed as quickly as possible about their exposures to each suspect item. Fading memories, the chance of obtaining still-available samples of implicated food or water, and the opportunity to obtain fecal specimens before the patient is started on the ubiquitous broad-spectrum antibiotics are all reasons for rapid response.

Case and control numbers do not have to be the same; the calculations compare ratios so equal numbers in each group are not needed. Generally a 1:1 to 1:2 ratio of cases to controls is perfectly adequate.

Where the Source Is Still Viable, Alert the Public of Potential Risks as You Become Aware of Them

As interview data from cases and controls are accumulated, leading to formation of hypotheses about the source of the illness, human resources should be deployed in two additional essential tasks: (1) tracking down and confirming the hypothesized source of the illness, and (2) promptly issuing warnings to all affected groups about the possible risks from any source that is still accessible, *with assurances that further bulletins will be issued as soon as confirmation is received.* This *precautionary principle* is a vital component of risk management in modern public health. Waiting for *absolute* confirmation before releasing warnings and advisories should not be an option in the twenty-first century. The principle holds that while false alarms can be quickly forgiven, further illness should be avoided at the highest priority. Failure to heed this step has contributed to needless suffering and severe damage to reputation, trust and credibility.

---

[1]A case-control approach is necessary because *unlike* the data in Table 5B, we rarely have full information about <u>all</u> the attendees, and therefore the true *incidence/attack rate* is not available. Very rarely, when *all* cases and controls are available for interview, we would have the *true incidence rates for ill and for not-ill persons* and this would allow a "*retrospective cohort study*" to be carried out. Under such circumstances, and using Table 6 as an illustration, we could state that of 33 persons exposed to item *X*, 25 persons (75.8%) had become ill compared to 12 ill of 34 not exposed (35.3%).

Use of Exposure Rates Rather Than Attack Rates

Where incubation times are longer than a day, there is increasing likelihood that only a small proportion of the non-ill people will be available for interview, and on many occasions, not even all the ill persons can be contacted. The point here is that the investigator is usually working with sub-sets of the true cases and controls. The 30 controls in Table 6 and possibly even the 37 cases may have been drawn from larger groups, and therefore we cannot state the *incidence rate*, for example, as: "...*25 of 33 exposed were ill*," because the "33" had been artificially assembled, and may not resemble the true incidence at all. We CAN, however, use *exposure rates*, for example: "...*of 37 Ill persons, 25 (67.6%) had been exposed to X*," and, "...*of 30 who were not-ill, only 8 (26.7%) had been exposed to X*." The overwhelming majority of waterborne or foodborne illness investigations are run as "case-control" studies (or to be more accurate, "case-comparison" studies, as very little true "controlling" is accomplished during the selection of the comparison group).

A *broadcasted* invitation to all who *might* have been exposed to come forward, typically results in few non-ill persons volunteering information, because non-affected individuals believe they have little if anything to contribute. This reduces validity even further, and more active recruitment is often necessary to convince them that their information is just as essential for the investigation as are the contributions from the less-fortunate attendees.

Odds Ratio as a Measure of Risk

Let us examine a waterborne illness suspected as being due to the consumption of water bottled from a certain spring. You have found 60 people who meet the case definition of illness, and another 29 non-ill people in same neighborhood who report no symptoms at all, and who will be your controls. In Table 7 we display the data and ask the question: "*is drinking this water related to the risk of illness?*" Whenever a 2×2 table appears, the first step is to calculate the odds ratio (OR).

An odds ratio tells us if there is a relationship (where $OR \neq 1$), and the *strength* of the relationship (the OR value itself). It also clearly indicates the *direction* of the relationship: *was* <u>*drinking*</u> *or* <u>*not-drinking*</u> *the dangerous activity?* This is easily

Table 7 Odds ratio

	Ill (cases)	Not-ill (controls)	Total	Label the four cells a, b, c, d as shown. The odds ratio is calculated by cross-multiplying
Drank spring water	56 a	14 b	70	$$\frac{(a \times d)}{(b \times c)} \text{ or } \frac{(56) \times (15)}{(14) \times (4)} = \frac{840}{56} = 15.0\ldots(\text{the odds ratio})$$
Did not drink water	4 c	15 d	19	This is interpreted as "*An ill person was 15 times as likely to have drunk spring water compared to a person who was not ill.*"
Total	60	29	89	

determined by finding the dominant pair from (a×d) or (b×c). In the example above, (a×d) is greater, so cell "a" links the row *drank* with the column "*ill*," while cell "d" links "*not drink*" with "*not-ill*." This assumption is not as obvious as it may seem; the cause of the illness may have been whatever "other" thing was drunk by those who avoided spring water!

It is important to clarify that the odds ratio yields *the strength of the association*, not the statistical significance. Most OR values (where many exposures are being assessed) will be close to 1.0 (= "*no association*"), while an OR *clearly exceeding 1.0* signifies a positive association between this exposure and illness, such that this exposure *increased the risk of illness*.

An OR < 1.0 is *protective*, meaning that exposure to this factor *reduced the risk of illness* compared to the other group. For example, an OR of 0.25 means that the exposed group had only *one-quarter* the risk of illness compared to the non-exposed group. The *non-exposed* group therefore has a *greater* risk (by a factor of 4). While this *protective* effect can be due to true therapeutic protection (e.g., exposure to antibiotics when you have an infection), it is frequently explained as "statistical" protection. As an example, consider an outbreak where everyone consumed only one of two possible types of bottled water. One source, A, contains a pathogen, and B does not. If the ill people were found to be five times more likely to have consumed type A (odds ratio=5.0) then the not-ill would have five times the rate of consuming water B, and only one-fifth of the rate of choosing water A (OR=0.20 or 20%). This can also be read as the *risk of illness for the non-exposed group*, or as the *risk of staying well* by the *exposed* group. An easier way to interpret an OR less than 1 is to place 1 *over* the OR to reveal a value greater than 1, but clearly labeled "protective."

Weakness and Strength of the Odds Ratio

The OR is a *ratio* between numbers, and therefore not sensitive to the actual numbers of people in individual cells, an important consideration when the numbers of subjects are relatively small. This is illustrated by the common question: "*How large does an odds ratio have to be before it is considered evidence of an association?*" A popular response is "*at least 2.0*," but this must be considered with extreme caution. For instance, with very large studies, an OR of 1.12 (barely more than 1.0) can be shown to be very highly significant statistically ($P=0.001$), whereas in a small-n study, an OR of even 5.0 may not achieve statistical significance.

The odds ratio is certainly a useful measurement, and should always be used when a 2×2 table is encountered. It will quickly advise you (1) that there is an association, (2) the strength of that association, and (3) the direction of the association, none of which are specifically measured by a test of statistical significance. Unfortunately, it is not reliable with small cell sizes, and is unable to answer the question: "*How likely is it that these numbers could happen just due to chance?*" For this, we need to test the statistical significance.

The best advice is to use the OR (or relative risk where appropriate) *together with* a test of statistical significance. Most online statistic calculators or laptop versions of SAS, SPSS, EpiInfo, etc. will give a selection of useful statistics (odds ratio, relative risk, several versions of chi-square, and Fisher's exact test, both one-tailed and two-tailed.)

Testing Statistical Significance

Basic Concepts

In keeping with all scientific enquiry, we begin by advancing the notion (the "*null hypothesis*") that there is *no association* between the exposure and the illness, and attempt to *support* that notion. If insufficient evidence is found to support the null hypothesis, we *reject* it and cautiously consider that an **association** may exist between the two variables. This can be described as a *statistically significant* association. Two methods of testing are presented: the **chi-square test** (written χ^2 and pronounced "ky"-square) for most 2×2 (or larger) tables, and the **Fisher's Exact Test** (only for 2×2 tables) when chi-square is not valid due to the numbers in the cells being too small (the following sections give advice about this decision).

The Chi-Squared Test (χ^2)

The *original data* value in each cell we call the *observed*, or "*O*" value, and these are compared with the numbers that you would *expect* ("*E*" values) if there were *NO* *relationship at all*; that is, if the variables were *not* related, and the data were arranged purely by chance (as stated by the null hypothesis). The chi-square test measures the difference between the *O* and *E* values. If they are close, we have to accept that there may be no real relationship; if far apart, we can reject the null hypothesis and cautiously declare that exposure and illness were probably related. Numerous online statistical calculators can be used to yield ORs, RRs, and chi-square values.[2] If you prefer to do the calculation by hand, construct a 2×2 table as shown, with "observed" data, marginal totals, and the grand total. The expected "*E*" values are found from:

$$E = \frac{(\text{row total})\times(\text{column total})}{\text{grand total}}$$

For cell "a": $E = \dfrac{70\times60}{89} = \dfrac{4200}{89} = 47.2$ The remaining "*E*" values are shown in parentheses.

---

[2] Epi-Info is a highly recommended suite of epidemiological and statistical programs, supported by the US CDC and WHO, and freely available for download in numerous languages. Full $2\times N$ table analysis is included.

Table 8 Chi-Square analysis for 2×2 tables

	Ill (cases)		Not ill (controls)	Total
Drank spring water	56 (47.2) a	b	14 (22.8)	70
Did not drink water	4 (12.8) c	d	15 (6.2)	19
Total	60		29	89

To make sure the chi-square analysis is appropriate for your table, you must be sure that all "E" numbers[3] are more than 5. The quickest way is to first calculate for the cell with the *smallest E* value; (this will be the cell *with the smaller column total* and *the smaller row total*.) In Table 8, the smallest E value will be cell "d," and this is calculated as $(19 \times 29)/89 = 6.2$. As this is >5, all other E values will be greater than this, so chi-sq. is valid. (Note that the smallest E value did not coincide with the smallest O value).

Chi-square (χ^2) is the SUM of $\dfrac{(O-E)^2}{E}$ for all four cells.

For cell "a" $\dfrac{(O-E)^2}{E} = \dfrac{(56-47.2)^2}{47.2} = \dfrac{(8.8)^2}{47.2} = 1.64$

For all four cells the sum (χ^2) is: $1.64 + 3.40 + 6.05 + 12.49 = 23.58$

An online statistical calculator will give you this same chi-square (χ^2) value. To verify by hand whether the O vs. E difference is statistically significant, compare your chi-sq. value (for a 2×2 table only) with 3.841. If your calculated value *exceeds* 3.841, then this is *unlikely to be due to chance*, and thus you can begin to believe that this exposure *did* influence the risk of illness, and you can reject the null hypothesis.

Statistical results usually include a probability (P) statement. This is the probability that the null hypothesis ("no association") is correct. The 3.841 value is the minimum needed for statistical significance, where the P is less than 5% ($P < 0.05$). Recall that the P is the *probability that NO real association exists between exposure and illness*. By convention, if $P > 0.05$ (more than 5%) then the relationship is declared **not statistically significant**. Where $P = 0.05$ or **<0.05**, then the relationship is **statistically significant**. The smaller the P value, ($P < 0.01$, $P < 0.001$, etc.) the more confidence you have that a relationship really exists. Other critical values exist for assessing calculated chi-sq. values, from larger tables than 2×2, and at more extreme levels of significance. A further chi-square calculation is shown as an appendix.

---

[3] The requirement is that not more than 20% of the cells should have an E value less than 5. In a 2×2 table, one cell (25%) already exceeds this. For larger tables (2×3, 3×3, etc.) the rule will allow one or more E values <5. Note that this applies to the E value, NOT the original O value in the cell.

Cell Size Limitation

Where a table greater than 2×2 is found to have more than 20% of the cells with an E value less than 5, chi-square is not valid. The solution is to *collapse either columns or rows* to allow the E values to increase. For example in Table 9A, two cells out of six (33%) have E values less than 5, but if *"high dose"* is merged with *"medium dose"* the resulting increase in observed (O) cell sizes is also reflected in greater E values, while the table becomes 2×2 (Table 9B). Some outcome information has been lost, but the chi-square analysis can proceed. If, after trying to collapse cells and/or rows, a 2×2 table is reached still with an E value <5, the Fisher's test is indicated.

Fisher's Exact Test: (Another Example is Shown as Form J2)

This procedure is reserved only for 2×2 tables where one or more expected (E) values is less than 5, making the chi-square test not valid. Our example is taken from an investigation into an outbreak of shigellosis presumed to be due to water from a well (Table 10). The odds ratio has been calculated as $(8 \times 6)/(4 \times 2) = 6.0$, *meaning ill persons were six times as likely to have drunk well water compared to non-ill*

Table 9 Collapsing rows or columns to obtain E values valid for chi-square analysis

(A) Before collapsing: chi-sq. not valid (two E values <5)				(B) After collapsing rows: chi-sq. now valid			
	Symptoms	No symptoms	Totals		Symptoms	No symptoms	Totals
High dose	16 (9.67)	9 (15.33)	25	"Any" dose	18 (12.37)	14 (19.63)	32
Medium dose	2 (2.71)*	5 (4.29)*	7	Control (no dose)	11 (16.63)	32 (26.37)	43
Control (no dose)	11 (16.63)	32 (26.37)	43	Totals	29	46	75
Totals	29	46	75	*Expected values <5 Expected values shown in parentheses			

Table 10 Fisher's exact test: Original data

	Ill	Not-ill	Total	
Drank well water	8 a	4 b	12 (a+b)	To test this we calculate a probability value (P) directly using $$P = \frac{(a+c)! \times (b+d)! \times (a+b)! \times (c+d)!}{a! \times b! \times c! \times d! \times (a+b+c+d)!}$$ The "!" denotes a factorial, meaning that number multiplied by the next smallest number, and so on down to 1. (e.g.: $6! = 720$) $$P_1 = \frac{10! \times 10! \times 12! \times 8!}{8! \times 4! \times 2! \times 6! \times 20!} = 0.075018$$
Did not drink well water	c 2	d 6	8 (c+d)	
Total	10 (a+c)	10 (b+d)	20 (a+b+c+d)	

Table 11 Fisher's exact test: Adjusted data

	Ill	Not-ill	Total	
Drank well water	8 9 a	4 3 b	12 (a+b)	The "dominant" a and d are each increased by 1, while b and c are both decreased by 1, keeping marginal totals unchanged.
Did not drink well water	c 2 1	d 6 7	8 (c+d)	Recalculating:
Total	10 (a+c)	10 (b+d)	20 (a+b+c+d)	$P_2 = \dfrac{10! \times 10! \times 12! \times 8!}{9! \times 3! \times 1! \times 7! \times 20!} = 0.009526$

Table 12 Fisher's exact test: Final data

	Ill	Not-ill	Total	
Drank well water	9 10 a	3 2 b	12 (a+b)	Recalculating: $P_3 = \dfrac{10! \times 10! \times 12! \times 8!}{10! \times 2! \times 0! \times 8! \times 20!} = 0.000357$
Did not drink well water	c 1 0	d 7 8	8 (c+d)	$P_{Total} = P_1 + P_2 + P_3$
Total	10 (a+c)	10 (b+d)	20 (a+b+c+d)	$P_{Total} = 0.0750 + 0.0095 + 0.0004 = 0.0849$

persons. An attempt to use chi-square is prevented by at least one E value less than of 5. (Cells c and d both show E values as $(8 \times 10)/20 = 4.0$). The starting null hypothesis is *"that no relationship exists."*

This is not quite the end of the calculation however. The goal is to calculate the probability of the *original data occurring plus all more extreme probabilities.* The original data have to be adjusted by increasing the "dominant" pair of cells by +1 and the others by −1, while leaving the margin totals the same (Table 11).

Because no zero has yet appeared in the matrix of cells, we continue to increase the "dominant" pair by +1 and obtain a zero. The next calculation is the last. (By convention, 1! and 0! = 1) (Table 12).

Interpretation: No reference table is required. The total calculated probability (0.085) is exactly the probability that the null hypothesis (*"that there was no relationship"*), is correct: 8.5%.[4] By convention, for a result to be significant statistically, that probability (P) must be less than 5% (<0.05), so in this instance we are *not* able to reject the null hypothesis and must conclude that the relationship *could* have occurred by chance alone more than 5% of the time. The odds ratio of 6.0 is explained as the number of times more likely it was for a shigellosis victim to have drunk well water than for a non-ill person. This increased risk would normally be impressive, but because of the small number of persons in the study, it has been

---

[4] The first probability (P1 = 0.075) was already in excess of 0.05, so it was already not significant, and further additions would increase this value still further. The calculations could therefore have stopped after the first probability, with the statement "P > 0.05, not-significant." The calculations here are carried out in full to illustrate the process of working toward a full and final probability.

found not to pass the test of statistical significance. A basic write up of the results might read:

"A relationship exists between drinking well water and developing shigellosis. A shigellosis patient is six times more likely to have drunk water from the well compared to a non-ill person. This relationship is not statistically significant, however, and could have occurred by chance alone more than 5% of the time. The null hypothesis of no-relationship cannot be rejected." [1 df, P > 0.05, OR: 6.0, not statistically significant.]

Summary Tables

With the odds ratio (OR) calculated for all the suspected exposures, and the chi-square test or Fisher's exact test calculated for the strongest of these, all the results can be displayed in a composite table.

Earlier protocols for the investigation of waterborne and foodborne diseases encouraged the use of the "***factor-specific attack rate table***" (for example the "*food-specific attack rate table*"), but where only a "convenience sample" of controls and cases are available, we are unable to derive valid incidence/attack rates. Investigators are discouraged from using it as it may produce misleading results. The ***exposure-rate table for cases and controls*** is preferred in all case-control studies, and compares the rates of exposure to each factor between both the ill and non-ill people.

Table 13 displays six exposure factors from a hypothetical outbreak involving water contamination. Exposure rates are calculated from both cases and controls. The "spring-water" data that we used for the odds ratio calculation example in Table 8 appears as the first exposure in Table 11. The column headed "Differences in exposure rates" subtracts the exposure rate among the non-ill from the exposure rate among the ill. [Use: Exp. rate (cases) minus Exp. rate (controls), keeping the signs correct]. You are looking for a large positive difference to indicate the most likely culprit. The spring water shows the largest positive difference at +45%. The odds ratio of 15.0 supports this, again the largest value, indicating *that ill persons were 15 times more likely to have drunk the spring water compared to non-ill persons in this group*. Hence both the large positive difference in exposure rates and the large odds ratio point to the spring water being the likely source of the illness, and it is certainly the strongest association between illness and any of the exposures shown. The chi-square value has also been added (23.6), as well as the associated *P* value. Taken together, the evidence clearly points to this factor as the culprit.

In those less-common circumstances in which ALL the ill and non-ill persons can be contacted for interview, the table can be rearranged to show *attack rates (incidence rates)* for each of the suspect factors (Table 14). Here, the column of "differences" shows the *attack rate (exposed) minus the attack rate (non-exposed)*, $(I_E - I_N)$, and again a large positive difference will point to the culprit. This measure is called the *attributable risk* and for the spring water example we obtain +59%, the largest value of all the risk factors. Also, because of the availability of valid attack rates (incidence rates), the *true relative risk* (RR) is available, and can be substituted

Table 13 Exposure-rate table for cases and controls (for use when data are a sample of both cases and controls)

Factor	Cases (ill)				Controls (not ill)				Diff. in exp rates (%)	Odds ratio[a]	Chi-Sq	P value[b]
	Exp (a)	Not exp (c)	Total (a+c)	Exp rate (a/(a+c)) (%)	Exp (b)	Not exp (d)	Total (b+d)	Exp rate (b/(b+d)) (%)				
Spring water	56	4	60	93	14	15	29	48	+45	15.0	23.6	<0.001
Soft drink	50	10	60	83	28	1	29	97	−14	0.18	3.15	0.08 ns
Water cress	48	12	60	80	23	6	29	73	+7	1.04		
Washed lettuce	35	25	60	58	16	13	29	55	+3	1.14		
Washed berries	51	9	60	85	24	5	29	83	+2	1.18		
Tomatoes	50	10	60	83	24	5	29	83	0	1.04		

[a]See Form J1 for an illustration of the odds ratio calculation

[b]The P (probability) value is a statement of statistical significance. It indicates the probability that there is NO relationship between the factor and the illness. So as the number becomes very small (as shown here) we can be increasingly satisfied that a real relationship does exist. See the Statistical Significance section for the correct way to calculate this

ns = not significant

Table 14 Attack-rate table (for use when data are available from ALL cases and ALL controls)

| Factor | For those EXPOSED (consumed item) | | | | For those NOT EXPOSED (did NOT consume item) | | | | | Diff. in attack rates (%) | Relative risk[a] | Chi-sq. | P value[b] |
	Ill (a)	Not ill (b)	Total (a+b)	Attack rate (a/(a+b)) (%)	Ill (c)	Not ill (d)	Total (c+d)	Attack rate (c/(c+d)) (%)				
Spring water	56	14	70	80	4	15	19	21	+59	3.80	23.6	<0.001
Soft drink	50	28	78	64	10	1	11	91	−27	0.71	3.15	0.08 ns
Water cress	48	23	71	68	12	6	18	67	+1	1.01		
Washed lettuce	35	16	51	69	25	13	38	66	+3	1.04		
Washed berries	51	24	75	68	9	5	14	64	+4	1.06		
Tomatoes	50	24	74	68	10	5	15	67	+1	1.01		

[a]The ("true") relative risk (also called the risk ratio) is only used when we have the true incidence data. It is calculated as the attack rate (exposed) over the attack rate (non-exposed), or I_E/I_N

[b]The P (probability) value is a statement of statistical significance. It indicates the probability that there is NO relationship between the factor and the illness. So as the number becomes very small (as shown here) we can be increasingly satisfied that a real relationship does exist. See the Statistical Significance section for the correct way to calculate this

ns = not significant

for the odds ratio. The data in Table 14 shows the same data as Table 13 rearranged for easy comparison. Values in the column of "differences" are not the same as for Table 13 and of course the relative risks are not the same as the odds ratios in Table 13, but both of these results still point clearly to the suspect exposure.

In both analyses, spring water is clearly the factor most strongly associated with illness. It is important to note that in both tables a high rate (exposure- or attack-) on the left side taken by itself is meaningless until it is compared with the rate from the right side of the table. This again underscores the importance of gathering complete data from the non-ill as well as the ill.

An interesting phenomenon is visible in the second factor listed (soft drink). The OR is listed as 0.18, which is "protective," meaning that this factor is strongly associated with NOT being ill. It is the equivalent of OR equal to 5.55 (1/0.18), and the chi-square is seen as quite large, although not enough for statistical significance. This is sometimes seen where TWO factors are "in competition" with each other; if everyone had drunk one item, and the spring water was the contaminated source, then those drinking the *other* item would be strongly "protected" because they did *not* drink the spring water, and this shows clearly. All other factors have OR values very close to 1.0.

Most attack rate tables record some persons who did not ingest the suspect vehicle but who nevertheless became ill. Plausible explanations are that (a) some people forget which beverages or foods they ingested; (b) some might have become ill from other causes; or (c) some may have exhibited symptoms with a psychosomatic rather than a physiological origin. It is also not unusual for the table to include some persons who ingested contaminated water or food but did not become ill. Plausible explanations are that (a) organisms or toxins are not always evenly distributed in water or food and consequently some persons ingest small doses or perhaps none at all; (b) some persons eat or drink larger quantities than others; (c) some are more resistant to illness than others, and (d) some will not admit that they became ill, or fail to report it.

Whichever table is used, the combined totals for cases (ill) and controls (well) are fixed and should not change for each exposure unless there are "missing" responses from interviewees.

While some procedure manuals include *confidence limits* around both RR and OR, this may be omitted here as the use of the chi-square test or Fishers Exact test yield the statistical significance for both tables.

Other Associations

Quantity-of-Water Ingested

Illness caused by ingestion of waterborne toxicants and some pathogenic organisms can be dose-related in that the risk of developing symptoms, and their severity varies with the quantity ingested. Where the suspect water is no longer available (for example, the well may have been quickly super-chlorinated to break the chain of infection before samples were taken), attack rates can be based on the amount of water usually drunk per day by each person. This is easily extended to other non-treated sources of water such as ice cubes, water-reconstituted fruit juices, and flavored crystals. A comparison of attack rates at various water intake levels may provide valuable evidence that water is, or is not, the vehicle responsible for the outbreak. For an example, see Table 15. Here, the entire group was 210 people and we have interviewed them all, so we are justified in calculating the attack/incidence rates:

In this example, the attack rate increased as the consumption of water increased, which suggests that the illness was directly related to water and the agent it contained. This is a trend established from the group as a whole, and an individual's experience may vary with factors such as (a) preferences of water ingestion, (b) intermittent contamination, (c) unequal distribution of the contaminant, or (d) varying susceptibility of individuals. These data can be compared with rates from persons who ingested no water, but only hot tea, hot coffee, soups, and/or other safe sources of liquids. If unheated water was indeed the vehicle, and the agent was a living biological agent, these persons should have attack rates showing no increase in risk of illness. (Outbreaks from a toxic agent may be unaffected by chlorination, boiling, and some types of filtering.)

The data can be displayed in a contingency table as follows for analysis using chi-square procedure (Table 16).

Table 15 Number of glasses of water and water/ice-containing beverages usually ingested per day by interviewees

	Ill	Not ill	Total	Attack rate (%)
5 or more	15	30	45	33
3 or 4	23	59	82	28
1 or 2	9	48	57	16
<1/day	2	24	26	8
Total	49	161	210	

Table 16 Data from Table 13 arranged for Chi-Square analysis

# glasses water/day	Ill	Not-ill	Total	
5 or more	a 15 (10.5)	b 30 (34.5)	45	Follow the procedure as for a 2×2 table. The original data are considered the "observed" (O) values, and we calculate the expected (E) values using: $\dfrac{(\text{row total}) \times (\text{column total})}{\text{grand total}}$
3–4	c 23 (19.1)	d 59 (62.9)	82	The expected numbers have been placed in parentheses. All E values are 5 or more, although this table could allow one E value that was less than 5[a]
1–2	e 9 (13.3)	f 48 (43.7)	57	The degrees of freedom (df) are calculated as $(\text{No. of rows} - 1) \times (\text{No. of columns} - 1),$ or $(4-1) \times (2-1) = 3$
<1/day	g 2 (6.1)	h 24 (19.9)	26	Chi-square is obtained by calculating … $\dfrac{(O - E)^2}{E}$
Total	49	161	210	… for each cell and adding the eight values obtained.

[a] For Chi-square to remain valid, not more than 20% of the cells can have an 'E' value less than 5. For 2×2 tables, a single cell is 25% of the total so such a table may have *no* cells with an E value less than 5. Tables 2×3 or 2×4 can still employ Chi-square with the E value of *one* cell less than 5

For this example, chi-square equals 8.90 and if calculated by computer or online, P will be shown as $P=0.031$. Reference to Form J1 confirms that for a 2×4 table (3 df), the calculated chi-square (8.90) exceeds the critical value for statistical significance at the 0.05 level (7.82), allowing us to claim statistical significance at $P<0.05$. Odds ratios are normally associated only with 2×2 tables, but here, the OR can usefully be calculated on selective cells or groups of cells as long as you clearly explain the selection process. For instance, persons who were ill were 2.8 times more likely to have drunk three or more glasses of water per day compared to those who were well. For this calculation we *collapse cells* into a 2×2 table and cross-multiply: $(a+c) \times (f+h)/(b+d) \times (e+g) = (38) \times (72)/(89) \times (11) = 2736/979 = 2.79$. Alternatively, because we have all people involved, we can compare the attack rates (AR) for each intake level, and observe the increasing attack rate as the intake increases: For five or more glasses/day, AR: 33%, for 3–4/day, AR: 28%, for 1–2/day, AR: 16%, and for <1/day, AR: 8%. We might summarize as follows:

"There was a relationship observed between the quantity of water consumed each day and the risk of illness. The incidence rate increased with the quantity consumed from 8% for <1 glasses/day to 33% for five or more glasses/day. This relationship is statistically significant. The null hypothesis of no association can be rejected." [Chi-square: 8.90, 3 df, P<0.05]

Other Water-Related Exposures

Water as a vehicle can deliver pathogenic organisms in many ways beyond simply drinking a glass of water, or using a drinking fountain. Investigators should be sure to ask about the preparation of ice-cubes, the mixing of fruit flavored crystal drinks, reconstituting concentrated orange juice, brushing and rinsing teeth, and washing hands, utensils, or containers. Swimming or playing in muddy pools or even swimming pools have caused waterborne poliomyelitis, and naegleriasis, while swimming in saltwater inlets have allowed inadvertent infections from *Vibrio parahaemolyticus* and *V. vulnificus*. Unwashed plastic jugs containing poster paint residue have caused rapid illness when drink crystals are reconstituted in them, while refillable plastic containers and bottles have a long history of contamination from biological and chemical agents. In the late 1970s, an increase of viral ear, nose, and throat infections among people who were using parkland next to a river was hypothesized to have been due to people waterskiing on the river and creating an aerosol. The river was the receiving body for effluent from a water treatment plant upstream.

Interpret Results from Water Samples

Record all laboratory results on Form I, *Laboratory Results Summary*. Compare epidemiological and statistical results with on-site observations, laboratory results and the information summarized on Form I. The agent responsible for the outbreak can be determined by (a) isolating and identifying pathogenic microorganisms from patients, (b) identifying the same strain and/or PFGE pattern or genetic sequence of pathogen in specimens from several patients, (c) finding toxic substances or substances indicative of pathological responses in specimens, or (d) demonstrating increased antibody titer in sera from patients whose clinical features are consistent with those known to be produced by the agent.

When implicating the water as a likely (or presumptive) vehicle of transmission, ideally identification of a pathogen in samples of suspect water will correspond to the one found in clinical specimens from ill persons or that produces an illness that is compatible with the incubation period and clinical features of the ill who were exposed to the water. For organisms that are common in the gastrointestinal tract or that have multiple strains, compare strains isolated from ill persons with strains isolated from the suspected water. Additionally, specific microbial markers (e.g., serotype, phage type, immunoblotting, plasmid analysis, antibiotic resistance patterns, restriction endonuclease analysis, nucleotide sequence analysis) or chemical markers identified by chromatography or spectrophotometry can be used for this purpose. For confirmation of water-related transmission, the same pathogen strains should be found in both the ill persons and the epidemiologically implicated water. However, due to the period of time that may have passed after the outbreak was actually reported, and to methodological issues, such as the need for concentrating pathogens in water samples, it is often unlikely that the outbreak-associated pathogen will be found in the water samples.

Laboratories frequently test water samples for indicator organisms, such as fecal coliforms, *Escherichia coli*, or enterococci, rather than pathogens. The finding of these bacteria in high densities in the water may indicate contamination (from a fecal source) and implicate the water was a possible vehicle. However, the finding of increased indicators in water samples alone is insufficient evidence to confirm the water as the source of an outbreak.

The probable source of contamination or the situation that allowed contamination to reach and survive in a water supply (e.g., water supply not disinfected or inadequately disinfected, inadequately filtered, or upstream to sewage or agricultural discharges; cross connection between sewerage and drinking water pipes; well improperly constructed; nearby septic tank system; or livestock in water supply) can often be identified, but the etiologic agent in the water may never be found. Success in finding the etiologic agent is most likely where (a) the incubation period of the illness is short, (b) the agent is stable in water and the system is static, or (c) large amounts of the agent are being continually added to the water supply. Try to recover and identify the specific agent whenever a water supply is suspected to be the vehicle of transmission, even if finding the etiologic agent is likely to be difficult and not considered practical for routine monitoring of water supplies. If water samples do not reveal a likely causal agent, clinical data as well as time, place, and person associations can cast strong suspicion on a water supply, particularly if indicator organisms are found in the water. Tests other than those for pathogens, however, are frequently used to evaluate water supplies on a routine basis.

Interpret Physical and Organoleptic Tests

Organoleptic tests attempt to evaluate the total effect of all compounds present in water that can be measured by the senses of taste, smell, or sight. Results cannot be expressed in terms of specific compounds present, and the measured qualities are usually a result of a mixture of compounds. These tests are often empirical and arbitrary, but changes in the physical qualities of water (such as pH, turbidity, color, odor, or taste) can indicate abnormalities of the water. Outbreaks have occurred, however, when turbidity readings have met present standards and when water appeared and tasted good.

Interpret Chemical Tests

Chemical examination of water is useful for (a) detecting pollution (especially from industrial wastes and pesticides), (b) determining effectiveness of treatment processes, (c) evaluating the previous history of the water, (d) determining hardness, and (e) detecting the presence of specific toxins. Results are usually expressed in milligrams per liter (mg/L=ppm, parts per million), or micrograms per liter (µg/L=ppb, parts per billion). Historically, acute water-related outbreaks seldom involve chemical substances, so chemical tests are not requested routinely unless either (a) circumstances indicate possible chemical contamination or (b) clinical symptoms suggest chemical poisoning.

Flowing water in a distribution system can be monitored to determine chlorine residual. Free available residual chlorine refers to that portion of the total residual chlorine remaining in chlorinated water at the end of a specific contact period that will react chemically and biologically as hypochlorous acid or hypochlorite ion. The reaction is influenced by pH and temperature. Total or combined residual chlorine refers to chlorine that has reacted with ammonia or other substances and is not available for further reactions, as well as the free available chlorine. A chlorine demand exists in a chlorinated water until a free available residual is produced. A free available chlorine residual, e.g., 1 mg/L (1 ppm) or higher, maintained throughout the distribution system of a community supply is an indicator of safety from enteric bacteria but not necessarily from pseudomonads, viruses or parasites. Outbreaks have occurred when chlorine residue levels have met present standards.

Interpret Microbiological Indicator Tests

Analyses for microbial indicator organisms provide information on the microbiological quality of water and guidance as to its safety for consumption or contact. Indicator organisms are easier to test for than pathogenic organisms, and some serve as a surrogate measure of fecal contamination in water. The absence of indicator organisms in the water, however, does not guarantee water safety; numerous outbreaks of water-related disease have occurred from water in which no indicator organisms were detected. Evaluation of the safety of water should be based upon a combination of results of (a) an on-site study to identify sources and modes of contamination and means by which contaminants survived treatment and (b) appropriate laboratory analyses. Microbiological results should be compatible with observed sources of contamination and/or treatment failures found during the investigation.

Heterotrophic Plate Count (HPC)

Although all natural waters contain bacteria, the number and kind vary greatly in different places and under different climatic and environmental conditions. The number of bacteria isolated and reported, however, often represents only a fraction of the total number present, for several reasons. Colonies seen on agar plates develop from either single organisms or clusters or chains of organisms. Heterotrophic bacteria represent only those that can use organic matter and grow at the selected temperature (30–35°C) within 48–72 hours under aerobic/microaerophilic conditions in/on a defined medium when the standard test (spread plate, membrane filter or pour plate) is used. The HPC may also be done using different media under different incubation times/conditions. (Higher counts are usually found when the longer incubation periods are used.) Also, certain microorganisms are unable to grow aerobically either in or on the medium used. Because of these variables, the terms Total Plate Count (TPC), Standard Plate Count (SPC) and Aerobic Plate Counts (APC) should not be used.

HPCs serve as an index of changing sanitary conditions. In general, counts of good-quality well water are fewer than 200–500 colonies per mL. Densities in surface water are higher, but quite variable, depending on water temperature, sources of pollution, amount of organic matter present, and soil that washes into the water. The sources of pathogens, toxic substances, or fecal contamination may not increase the HPC of a surface water sample as much as washings from soil. Nevertheless, marked changes in the number or kind of microorganisms should be viewed with concern, at least until the reason for the change is discovered. Heterotrophic plate counts greater than 1000/mL and some specific antagonistic species may interfere with the growth or recovery of pathogenic or indicator organisms. Some heterotrophic species are opportunistic pathogens that may pose a health threat to immunocompromised persons.

Total Coliforms

The coliform group of bacteria comprises those from non-fecal environmental sources, and those from animal and human intestines, including *Escherichia coli*. The environmental species of non-fecal bacteria are found in soil, on fruits, leaves, and grains, and in run-off water, especially after heavy rains. Some of these species are capable of surviving in water longer than *E. coli*. Furthermore, some coliform strains and can multiply on decaying vegetation in water, in biofilms in pipelines, or on pump packings, washers, and similar materials. Therefore, finding coliforms may not be indicative of fecal contamination, although most water utilities have standards for coliforms in water. Fecal coliforms are present in large densities in all human and animal feces, normally much higher than pathogens which are typically only present in infected persons and normally at lower levels. As such, high populations of fecal coliforms can indicate recent sewage pollution of water, but are not always indicative of pathogens present, particularly viruses and parasites. None of the coliform group, however persists as long as most viral or protozoan pathogens in water, and indicator bacteria described below (fecal streptococci and *Clostridium perfringens*).

Typical chlorination or ozonation of water inactivates coliform bacteria. Presence of the coliform group or even a high population of coliform bacteria is not proof that a treated water supply contains pathogens. However, coliforms can provide a warning that either the water treatment was inadequate or contamination occurred after treatment, and that some pathogens may be present. As mentioned above, under some conditions, pathogens may be present where there are few or no coliforms. Furthermore, unlike coliforms, many parasites and viruses are resistant to normal levels of disinfectants. Coliforms have little or no correlation with the presence of parasitic protozoa or pathogenic viruses.

The standard test for the coliform group may be carried out by a membrane filtration technique, a multiple-tube fermentation technique (presumptive test, confirmed test, or completed test), or a presence-absence test. Results of the membrane filtration technique are reported as colony forming units (CFU) per 100 mL of water. Results of the multiple-tube fermentation technique are reported as the most probable number (MPN) per 100 mL of water. This is a statistical estimation of the total number present, but the actual number can fall within a considerable range. Counts derived from these two methods are not necessarily the same, but they have the same sanitary significance. False-negative or false-positive results can also occur with the membrane filtration technique because of interfering background growth of nonfecal microorganisms.

Results of the presumptive test of the multiple-tube fermentation technique can be misleading, because other microorganisms frequently found in water also produce gas in laboratory media, and may thereby give false-positive results. Also, especially in waters containing a large number of microorganisms, some coliforms present may produce gas slowly, leading to false-negative results. The presence of coliform bacteria is corroborated by means of the second phase of the multiple-tube fermentation technique, known as the confirmed test. Positive results are usually considered confirmation of the presence of coliforms. A third phase of this test, known as a completed test, further ensures the correct identification of coliform bacteria.

A simple modification of the coliform test is to analyze for the presence or absence of coliforms in a 100-mL drinking water sample. The "presence-absence (P/A) coliform test" allows for simple examination of a larger number of samples.

When a positive sample is detected, it is advisable to measure coliform densities in repeat samples by one of the other methods to determine the magnitude of the contamination.

Thermotolerant coliform (fecal coliform). Coliform bacteria will frequently grow at a relatively high temperature, 44.5°C, unlike species or strains normally encountered in the environment, which usually have an optimal temperature near 30°C. This thermotolerant characteristic has been used in an attempt to separate coliform bacteria into those of so-called fecal and non-fecal origin. This test may provide better indication of fecal contamination than the coliform test, but it is however, unreliable. Positive results are not proof that either organisms of fecal origin or pathogens are present. The number of thermotolerant coliforms is considerably lower than the number of total coliforms in contaminated water; therefore, the test is less sensitive for testing treated drinking water. Furthermore, *Escherichia coli* O157:H7, which has been implicated as causing water-related illness, does not grow well at 44.5°C.

Escherichia coli. *E. coli* is common in feces of human beings, other mammals, and birds. It can also be found to grow naturally in the environment, specifically in tropical waters. Comprised of the larger coliform group, its detection in water is a more definitive indicator of fecal contamination, compared to total or fecal coliforms. However, a positive test result does not identify if the fecal source is human or nonhuman. Rather, the finding of *E. coli* in water serves as an indicator that fecal matter reached the water and provides a warning, but not proof, that pathogenic organisms may also be present. It should be noted that some strains of *E. coli* are pathogenic (see Table B).

Simple commercial P/A and quantitative tests have been developed to detect the presence of total coliforms and *E. coli* in 24 hours by observing color changes and fluorescence of the media under daylight and UV light. Such tests may be useful for field evaluation of microbiological water quality.

Enterococci (Fecal streptococci). Another group of organisms, collectively known as fecal streptococci, is also used as an indicator of fecal contamination. Enterococci (*Enterococcus faecalis*, *Enterococcus faecium*) are particularly used for testing recreational waters. Like coliforms, enterococci are normal inhabitants of the intestinal tract of human beings and other animals. In human feces, they occur in considerably lower numbers than *E. coli*. Some members of the group, such as *E. faecalis*, subsp. liquefaciens, however, have been associated with vegetation, insects, and certain types of soils. Enterococci generally survive longer than coliforms in fresh water, and therefore the source of contamination may be distant in either time or place from the site where samples were obtained. Their resistance is, however, less than that of *Clostridium perfringens*, enteric viruses, and parasites.

Like *E. coli*, simple commercial P/A and quantitative tests have been developed to detect the presence of enterococci in 24 hours by observing color changes under UV light, which may be useful for field evaluation of microbiological water quality.

***Clostridium perfringens* (sulfite reducing clostridia).** *C. perfringens* is also of fecal origin, but it occurs in feces in much lower densities than *E. coli* and can also be found in soils. Being a spore-former, it can survive for long durations in soil and water, and persist when all other bacteria of fecal origin have disappeared. Therefore, it is a useful indicator of remote or intermittent contamination in wells that are not frequently examined by the coliform test; but, it is not, by itself, evidence of recent contamination. Chlorine, in the concentration typically used in water treatment, does not inactivate all spores; and thus *C. perfringens* is not valuable in assessing the efficiency of chlorination for bacterial vegetative cells. Its long persistence and its resistance to chlorine make this organism a potential indicator for viral and parasitic organisms that have similar resistance and disinfectant susceptibility.

Coliphage. Coliphages, which are viruses that infect *E. coli*, are simpler to detect and enumerate, compared to other viruses, and are generally associated with fecal contamination. They have been considered as possible indicators of treatment effectiveness for human enteric viruses. Coliphages are categorized into two groups: the somatic phages, which enter *E. coli* via the cell wall and the male-specific phages, which enter *E. coli* through the sex pili. The somatic and male-specific phages are common in sewage and the feces of human beings and other animals, but in lower densities than the common fecal indicator bacteria, fecal coliforms, *E. coli*, and enterococci. Some strains appear to be more resistant to chemical disinfection than water-related pathogens or indicator bacteria.

Local Standards

Be aware of local standards for water distribution systems, private water systems, and recreational water. Although drinking water standards, such as the total number of coliforms allowed in a water sample, vary from jurisdiction-to-jurisdiction, it is generally agreed that any fecal contamination (e.g. fecal coliforms, *Escherichia coli*) render the water unacceptable for human consumption and may close down recreational bathing waters.

Interpret Tests for Pathogens

There are numerous pathogens that can be transmitted by water, many of which are also able to cause respiratory symptoms, in addition to the classical gastrointestinal symptoms. For a comprehensive summary of waterborne pathogens see "American Waterworks Association Manual of Water Supply Practices, M48 Waterborne Pathogens, 2nd edition (2006)."

For several reasons, analyses for pathogens are not usually conducted during routine water testing, or are only conducted by specialized laboratories. First, tests for pathogens are pathogen-specific, expensive, and often difficult to perform because they may require specialized trained personnel. Secondly, the etiologic agent of the outbreak is often unknown at the time of analysis; hence, many analyses would have to be done blindly. Thirdly, pathogens are not always recovered because they are heterogeneously dispersed and diluted in the environment, and their numbers decline in water over time. As a result, they may be absent or present in low densities by the time samples are collected following an outbreak. Fourthly, recovery efficiencies are often poor because microorganisms are stressed by disinfectants or the method is sensitive to interferences from the source waters environment, hence not easily recovered by routine methods. Additionally, recovery efficiencies for viruses and protozoa may be poor because of the interferences of substances within the sample matrix with method reagents (concentrating 1800 L of water down to 200 µL will also concentrate inhibitory chemicals and substances). Finally, the time required for isolation and identification is often long, and the number of samples is usually too small to allow the investigator to have much statistical confidence in the results when pathogens are not found.

Negative results should be reported as "Not Detected" because they do not ensure that the water sampled was not the source of the pathogen. Procedures used for many bacterial pathogens are qualitative because enrichment procedures are used. Quantitative procedures (e.g., MPN) require considerable work and are less reliable than those used for coliforms because small populations may be present, and these may be unevenly distributed. Despite these difficulties, pathogens that cause a syndrome similar to the one being investigated should be sought. See Tables B, C, and D, for descriptions of the disease syndrome associated with the pathogens described in the following material. Finding the same pathogen in specimens from patients and in water samples confirms water as a vehicle.

Submit Report

Summarize investigative data in a narrative report. Describe in this report situations that led to contamination of the water and survival of etiologic agents up to the time of consumption. Include all events that contributed to the outbreak to guide control and preventive measures. Compare your data with the listings in Table G (Guidelines for confirmation of waterborne outbreaks) and Table H (Guidelines for confirmation of water responsible for illness), and criteria for confirmation of vehicle responsible for waterborne illness before assigning the etiologic agent and the vehicle. Outbreak confirmation is based on (a) time, place, person associations, (b) recovery of etiologic agents from clinical specimens from cases and samples of water, and (c) identification of sources and modes of contamination and means by which pathogens or toxic substances survived treatment. All three of these, however, might not be found in any one investigation.

Complete Form K (Waterborne illness summary report). Attach the narrative and the epidemic curve. Also attach Form D2 (Case history summaries: Water/Laboratory data), all applicable parts of Forms G, Forms H, Form I, and other data that will provide supplemental information to reviewers.

Send this report through administrative channels to the appropriate agency responsible for waterborne disease surveillance. Make the final report as complete as possible, so that the agency can accurately interpret the results and develop a meaningful waterborne disease data bank. In the interest of continuing cooperation, give all participants in the investigation due credit and send each a copy of the report. Also, send copies of the report through administrative channels to agencies (a) that have jurisdiction over the implicated water, (b) that initiated the alert, and (c) that participated in the investigation.

Those concerned with water sources, treatment and recreation, as well as with public health, should make every effort to ensure the complete investigation and reporting of waterborne diseases. Without reliable, complete information, trends in waterborne disease incidence and causal factors of the disease are difficult to determine. Good surveillance is essential for detecting and evaluating new waterborne disease hazards.

Use Outbreak Data for Prevention

The primary purposes of a waterborne disease investigation are to identify the cause, establish control measures, and take actions to prevent future illness. Prudence may require some action before all the hypotheses regarding the water supply involved and the source of contamination are confirmed. Frequently the local health authority will issue a Boil Water Advisory if a microorganism is suspected to have contaminated the water. Refer to "Possible Precautionary Control Actions" section for a discussion of these precautionary control measures. If these measures have not already been considered, consider them now. Once control measures have been implemented, continue to monitor for disease to evaluate whether the measures were effective. In a waterborne event in Sydney, Australia (see Box 1) Sydney Water severely overestimated levels of *Cryptosporidium* and *Giardia* present in the water raising public alarm. Boil water advisories were announced and rescinded several times. However, it is better to announce boil-water advisories than to have thousands ill, as has happened in the past, such as the *Cryptosporidium* outbreak in Milwaukee in 1993.

Deficiencies in treatment must be corrected and defective parts of distribution systems must be repaired, beginning with those that either contributed to or had a high potential for contributing to the outbreak. The effectiveness of these efforts will be directly related to the thoroughness of the investigation. Document the source and the manner of contamination and survival of the etiologic agent through the water treatment process. Provide clear documentation of contributory factors, so that preventive measures taken will be specific to the problem.

If previous sanitary surveys have revealed, or if subsequent ones reveal, that conditions which contributed to the outbreak are widespread, initiate a training and education program. These programs can be developed for water treatment plant or recreational water operators and employees, engineers, homeowners, or other appropriate groups. Impress upon them the importance of proper construction and operation of facilities and proper protection, treatment, storage, and distribution of water. Follow up with periodic inspections and surveys and verify by sampling, as appropriate, to determine whether faulty conditions have been corrected or allowed to be reintroduced. Legal action may be necessary to ensure compliance with official standards and accepted sanitary practices.

Formulate solutions to problems found during outbreak investigations, and incorporate these into regulations for drinking, agricultural, industrial, domestic, and recreational waters. Inform the public, through mass media and other means available to your agency, of hazardous conditions that can affect their water supply, but do so only after hypotheses are confirmed. The public must be told of any potential or actual harm that may result from ingesting or contacting contaminated water and must also be informed of measures that they can take and that official agencies are taking to correct these conditions. The water supply and recreational water facilities must be verified periodically to determine whether critical processes are being monitored and operated within limits of appropriate public health standards (See Box 2, The Walkerton Outbreak).

Most waterborne illnesses are preventable, but prevention requires that those in the water treatment industry and in health and water-protection regulatory agencies be constantly vigilant to ensure that the hazards are understood and that questionable water treatment or delivery system construction or practices are avoided.

Acknowledgments The Committee and Association thank and cite the following persons for their assistance in critically reviewing parts of this edition:
E. Rickamer Hoover (CDC/ONDIEH/NCEH)
Richard Sakaji (East Bay Municipal Utility District, Oakland California)
John Hanlin (Ecolab, Eagan, Minnesota)
Phyllis Posy (Atlantium Technologies, Israel)

The Committee and Association thank and cite the following Committee members for their We would also like to acknowledge the contributors to the second edition of this manual:
Frank L. Bryan, O.D. (Pete) Cook, Kim Fox, John J. Guzewich, Dennis Juranek, Daniel Maxson, Christine Moe, Richard C. Swanson, Ewen C.D. Todd

The Committee and Association thank and cite the following persons for their assistance in developing, writing, editing, and/or critically reviewing the second edition of this manual:
Procedures to Investigate Waterborne Illness
Ruth A. Bryan, Robert Burhans, Rebecca Calderon, James D. Decker, John M. Dunn, David Frederickson, Arie H. Havelaar, Howard Hutchings, M. Louise Martin, George K. Morris, Dale Morse, J. Virgil Peavy, Patricia Potter, Richard Vogt, Irving Weitzman

Committee members of the first edition who established the objectives and scope of the manual and also developed some of the technical content that is included in this revision are gratefully acknowledged:
Herbert W. Anderson, K.J. Baker, Gunther F. Craun, Ward Duel, Keith H. Lewis, Thomas W. McKinley, R. Ashley Robinson, Richard C. Swanson, Ewen C.D. Todd

Further Reading

American Public Health Association. Control of Communicable Diseases Manual, 20th (ed.) DA Heymann. APHA Press. December 2014.
American Public Health Association. Heymann DL (ed). Control of Communicable Diseases in Man. 19th Edition. 2008.
American Water Works Association. Emergency Planning for Water Utilities, Fourth Edition. 2001.

American Water Works Association, American Society of Civil Engineers. Eds. Randtke SJ, Horsley MB. Water Treatment Plant Design, Fifth Edition. McGraw-Hill. 2012.

American Waterworks Association. Manual of Water Supply Practices, M48 Waterborne Pathogens, Second Edition, 2006. Denver, Colorado.

Baum R, Bartram J, Hrudey S. The Flint water crisis confirms that U.S. drinking water needs improved risk management. Environ. Sci. Technol. May 17, 2016. Accessed on: May 24, 2016 at: http:pubs.acs.org/doi/abs/10.1021/acs.est.6b02238

Bohm SB, Produce Safety- What's Going on Here? National Environmental Health Association NEHA-CERT EP0704 June 21 2007

Centers for Disease Control and Prevention. Atlanta, GA. Global Health - Division of Parasitic Diseases and Malaria Page last reviewed April 21, 2015. Accessed on June 2, 2015 at: http://www.cdc.gov/parasites/

Centers for Disease Control and Prevention. Atlanta, GA. Division of Foodborne, Waterborne, and Environmental Diseases. Outbreak Investigations. Page last reviewed on Sept 16, 2014. Accessed June 2, 2015 at: http://www.cdc.gov/ncezid/dfwed/waterborne/investigations.html

Centers for Disease Control and Prevention. Atlanta, GA. Reporting and Surveillance for Norovirus: CaliciNet. Page last updated on September 17, 2015. Accessed on February 19, 2016 at: http://www.cdc.gov/norovirus/reporting/calicinet/

Centers for Disease Control and Prevention. Atlanta, GA. National Outbreak Reporting System (NORS). Page last updated November 18, 2015. Accessed on February 19, 2016 at: http://www.cdc.gov/nors/

Centers for Disease Control and Prevention. Atlanta, GA. Surveillance Reports for Recreational Water-associated Disease & Outbreaks. Page last reviewed on June 2, 2015. Accessed on June 2, 2015. http://www.cdc.gov/healthywater/surveillance/rec-water-surveillance-reports.html

Centers for Disease Control and Prevention. Atlanta, GA. The Safe Water System. Effect of Chlorination on Inactivating Selected Pathogens. Page last reviewed on May 1, 2014. Accessed on August 14, 2015. http://www.cdc.gov/safewater/effectiveness-on-pathogens.html

Cox P, Fisher I, Kastl G, Jegatheesan V, Warnecke M, Angles M, Bustamante H, Chiffings T, Hawkins PR. Sydney 1998 - Lessons from a Drinking Water Crisis AWWA 95(5);147-161. 2003. Accessed on February 22, 2016 at: http://www.awwa.org/publications/journal-awwa/abstract/articleid/14804.aspx.

DiCaprio E, Ma Y, Purgianto A, Hughes J, Li J. Internalization and dissemination of human norovirus and animal caliciviruses in hydroponically grown romaine lettuce. Appl. Environ. Microbiol. 78(17):6143-6152. 2012.

Environmental Protection Agency Technical Guidance Manual LT1ESWTR Disinfection Profiling and Benchmarking. March 2003. Accessed on April 30, 2016 at: http://nepis.epa.gov/Exe/ZyNET.EXE?ZyActionL=Register&User=anonymous&Password=anonymous&Client=EPA&Init=1%3E%3Ctitle%3EEPA%20-%20Home%20Page%20for%20the%20Search%20site%3C/title%3E%3Clink%20rel=.

Environmental Protection Agency Guidance Manual for the Compliance with Filtration and Disinfection Requirements Public Water Systems Using Surface Water Sources. March 1991. Accessed on April 30, 2016 at: http://nepis.epa.gov/Exe/ZyNET.EXE?ZyActionL=Register&User=anonymous&Password=anonymous&Client=EPA&Init=1%3E%3Ctitle%3EEPA%20-%20Home%20Page%20for%20the%20Search%20site%3C/title%3E%3Clink%20rel=.

Environmental Protection Agency. Manual of Individual and Non-Public Water Supply Systems. EPA No. 570991004. 1991. Accessed on April 30, 2016: http://nepis.epa.gov/Exe/ZyNET.EXE?ZyActionL=Register&User=anonymous&Password=anonymous&Client=EPA&Init=1%3E%3Ctitle%3EEPA%20-%20Home%20Page%20for%20the%20Search%20site%3C/title%3E%3Clink%20rel=

Environmental Protection Agency. Manual of Small Public Water Supply Systems. EPA No. 570991003. 1991. Accessed on April 30, 2016: http://nepis.epa.gov/Exe/ZyNET.EXE?ZyActionL=Register&User=anonymous&Password=anonymous&Client=EPA&Init=1%3E%3Ctitle%3EEPA%20-%20Home%20Page%20for%20the%20Search%20site%3C/title%3E%3Clink%20rel=

Environmental Protection Agency (Federal Register, June 29, 1989, 40 CFR, Parts 141 and 142) http://nepis.epa.gov/Exe/ZyNET.EXE?ZyActionL=Register&User=anonymous&Password=an onymous&Client=EPA&Init=1%3E%3Ctitle%3EEPA%20-%20Home%20Page%20for%20 the%20Search%20site%3C/title%3E%3Clink%20rel=. Accessed on April 30, 2016

Hanley R, The Water Research Centre. Water Treatment Manual: Disinfection. Environmental Protection Agency, Wexford, Ireland. 2011. Accessed on February 24, 2016 at: https://www.epa.ie/pubs/advice/drinkingwater/Disinfection2_web.pdf.

Health Canada. Guidelines for Canadian Drinking Water Quality-Summary Table. Water and Air Quality Bureau, Healthy Environments and Consumer Safety Branch, Health Canada, Ottawa, Ontario. 2014. Accessed on February 24, 2016 at: http://www.hc-sc.gc.ca/ewh-semt/pubs/ water-eau/sum_guide-res_recom/index-eng.php.

Hipel KW, Zhao NZ, Kilgour DM. Risk analysis of the Walkerton drinking water crisis , Canadian Water Resources. 28(3); 395-419. 2003. Accessed on February 24, 1016 at: http://dx.doi. org/10.4296/cwrj2803395.

Hilborn ED, Roberts, VA, Backer L, DeConno E, Egan J, Hyde J, et al. Algal Bloom-Associated Disease Outbreaks Among Users of Freshwater Lakes- US 2009-2010, MMWR 63(1):11-15 Accessed on June 2, 2015 at: http://www.cdc.gov/mmwr/preview/mmwrhtml/mm6301a3.htm

Hoff JC, Akin EW. Microbial resistance to disinfectants: mechanisms and significance. Environ Health Perspect. 1986 Nov; 69: 7–13. http://www.ncbi.nlm.nih.gov/pmc/articles/PMC1474323/?page=1

Hrudey SE, Hrudey EJ. Safe Drinking Water: Lessons from recent outbreaks in affluent nations. London, UK: IWA Publishing; 2004

Hrudey SE, Hrudey EJ, Pollard SJ. Risk management for assuring safe drinking water. Environ Int. 2006 Dec;32(8):948-57.

Hrudey SE, Hrudey EJ. Ensuring Safe Drinking Water: Learning from Frontline Experience with Contamination. American Water Works Association. 2015.

Koreivienė J, Anne O, Kasperovičienė J, Burškytė V. Cyanotoxin management and human health risk mitigation in recreational waters. Environ Monit Assess (2014) 186:4443–4459.

Mac Kenzie WR, Hoxie NJ, Proctor ME, Gradus MS, Blair KA, Peterson DE, Kazmierczak JJ, Addiss DG, Fox KR, Rose JB, Davis JP. A massive outbreak in Milwaukee of Cryptosporidium infection transmitted through the public water supply. N Engl J Med. 1994; 331:161–167.

McLaren C., Colbourne J, Scott, R. Private Water Supplies: Technical Manual. 2016. Accessed on February 19, 2016 at: http://www.privatewatersupplies.gov.uk/private_water/files/Full%20Doc.pdf.

Mosse P, and Murray, B. Good Practice Guide to the Operation of Drinking Water Supply Systems for the Management of Microbial Risk: Research Project 1074. Water Research Australia, Adelaide, South Australia. 2015.

Murayama S, Mizawa M, Takegami Y, Makino T, Shimizu T. Two cases of keratosis follicularis squamosa (Dohi) caused by swimsuit friction. Eur J Dermatol. 2013 Apr 1;23(2):230-2.

O'Connor DR. Part one report of the Walkerton inquiry: The events of May 2000 and related Issues. Toronto, ON: Ontario Ministry of the Attorney General. 2002. Available from: Accessed on April 30, 2016 at: http://www.archives.gov.on.ca/en/e_records/walkerton/index.html

O'Connor DR. Part two report of the Walkerton inquiry: A strategy for safe drinking water. Toronto, ON: Ontario Ministry of the Attorney General. 2002 Accessed on April 30, 2016 at: http://www.archives.gov.on.ca/en/e_records/walkerton/index.html

Ontario Agency for Health Protection and Promotion (Public Health Ontario). Public health inspector's guide to the principles and practices of environmental microbiology. 4th ed. Toronto, ON: Queen's Printer for Ontario; 2013. Accessed on June 2, 2015 at: http://www. publichealthontario.ca/en/eRepository/Public_Health_Inspectors_Guide_2013.pdf

Schaefer P, Baugh RF. Acute otitis externa: an update. Am Fam Physician. 2012 Dec 1;86(11): 1055-61.

Sekhar M, Dugan A. Collect representative distribution system samples. Opflow (AWWA) 35(1): 20-23.

Sinisi, L., and Aertgeerts, R. (eds). Guidance on Water Supply and Sanitation in Extreme Weather Events. WHO Regional Office for Europe, Copenhagen, Denmark. 2010. Accessed on April 30, 2016 at: http://www.euro.who.int/__data/assets/pdf_file/0016/160018/WHOGuidanceFVLR.pdf.

Solomon EB, Potenski CJ, Matthews KR. Effect of irrigation method on transmission to and persistence of *Escherichia coli* O157:H7 on lettuce. J Food Prot. 65(4), 673-676. 2002.

Solomon EB, Yaron S, Matthews KR. Transmission of *Escherichia coli* O157:H7 from contaminated manure and irrigation water to lettuce plant tissue and its subsequent internalization. Appl. Environ. Microbiol. 68(1):397-400, 2002.

Appendices

Table A Equipment useful for investigations[a]

Item	Examples
Investigation guidelines and investigative forms	IAFP manual, "Procedures to Investigate Waterborne Illness, 3rd ed"; 50 copies of Form C; one dozen copies each of Forms E and F; two copies of form D and all parts of Form G, Epi-Info software (CDC, Atlanta).
Sterile sample containers	Water sample bottles (bottles for chlorinated water should contain enough sodium thiosulfate to provide a concentration of 100 mg of this compound per L of sample), plastic bags (Whirl-Pak® type), 250 mL, 1-L and 1-gal sized jars and jugs.
Sterile and wrapped sampling implements	Moore swabs (compact pads of gauze made from strips 120 cm [4 ft] by 15 cm [6 in.] tied in the center with a long, stout twine or wire—for sewer drain, stream or pipeline samples), fiberglass-epoxy bacterial filter cartridge, 0.3 μm; tongs, scoop or similar utensils for collecting ice.
Specimen-collecting equipment (for human specimens from cases and controls)	Sterile containers (with lids) for stool specimens, bottles containing a bacterial preservative and transport medium, mailer tubes or styrofoam box, sterile swabs, rectal swab units, tubes of bacterial transport medium, stool preservative medium for parasites, phlebotomy supplies for blood specimens.
Kits for testing chemical disinfectants and pH	DPD (*N,N*-diethyl-*p*-phenylenediamine) chlorine comparator with color disc for chlorine (0.1 ppm) and chlorine test papers; field-type pH meter or pH comparator with color disc or pH test papers; applicable pH indicator solutions and DPD reagent solution; dissolved oxygen testing unit.
Dye tracing study equipment	Fluorescein (yellow-green fluorescent) dye in powder form (10 packages containing 300 g each), in tablet form (100 tablets), or in liquid form (prepared by mixing 300 g in 1 L of water); fluorometer; filters (primary and secondary) for use with fluorometer; sample holder for continuous sampling or individual sampling; fluorometer recorder.
Disinfectant and neutralizer	0.5% w/v solution of calcium hypochlorite or 5.25% household liquid bleach; 50% w/v sodium thiosulfate.
Virus filtration equipment for viruses and parasites[b]	Large plastic container for storing water sample prior to concentration; portable electric or gasoline powered water pump with quick disconnect brass or stainless steel plumbing adapters or hose couplings; two filter holders for 10-in. water filter cartridges fitted for adapters or couplings; portable water meter fitted for adapters or couplings; four lengths of fiber-reinforced garden hose fitted with adapters or couplings; one length of a strong-walled supply hose fitted with adapters or couplings; 10-in. prefilter (3 μm nominal porosity wound polypropylene yarn filter with hollow perforated stainless steel core) cartridge filter; 10-in. virus absorbing filter pleated 0.2 μm porosity nylon membrane type (positively charged) for waters of pH values up to 8.5, or pleated 0.45 μm porosity glass fiber membrane type (positively charged) for waters of pH value of 7.5 or lower (e.g., Virosorb, 1-MDS, AMF/Cuno Meriden, pleated, 0.45 μm, glass filter); 1600 mL sterile, pH 7, 3% beef extract solution in 1 gal wide-mouth screw capped autoclavable polypropylene container for each sample to be collected; stands to support filter holders during filtration; for parasites 10-in. polypropylene yarn-wound cartridge filter, 1.0 μm porosity (e.g., Micro Wynd II™, AMF/Cuno; Meriden, CT. 1.0 μm normal porosity).

(continued)

Table A (continued)

Item	Examples
Supporting equipment	Laptop or tablet, with software; thermocouples of varying lengths with either recording potentiometer, data logger, or digital indicator; devices to take samples below surface and sediment samples; chemical smoke kit and/or micromanometer; Occupational Safety and Health Administration (OSHA) or equivalent approved respirator; sterile plastic gloves; plastic container liners for ice; waterproof marking pens; waterproof test tube rack; pencils, note pad; roll of adhesive or masking tape; labels; waterproof cardboard tags with eyelets and wire ties; flashlight; matches; test tube rack to fit tubes used; insulated chest or styrofoam container; packing material; camera with flash; spare batteries for all equipment; 95% ethyl alcohol; propane torch; refrigerant in plastic bags, liquid in cans, rubber or heavy plastic bags that can be filled with water and frozen; heavy-duty bags for ice, "canned ice," or cold-packs (blue ice).

[a]Assemble a kit to be kept in the agency responsible for investigating waterborne illness. It should include at least ten water sample bottles; ten 1-L, or gal jars or jugs; ten specimen collection containers or devices; and one each of the following supporting equipment and sterilizing equipment. Date of sterilization should be marked. Periodic resterilization or replacement of sterile supplies, media, or transport media is required to maintain the kit in a ready-to-use condition

[b]Similar equipment for sampling for either viruses or parasites may be available from national water, environmental, or health agencies

Table B Illness acquired by ingestion of contaminated water: a condensed classification by symptoms, incubation periods, and types of agents

Illness	Agent	Incubation or latency period[a]	Signs and symptoms[a]	Source of contaminated water	Specimens to collect	Factors contributing to outbreaks
UPPER GASTROINTESTINAL SIGNS AND SYMPTOMS [NAUSEA, VOMITING] PREDOMINATE Incubation period usually less than 1 h *Chemicals*						
Arsenic poisoning, acute (see also below CHRONIC ILLNESSES)	Arsenic (inorganic)	A few hours	Nausea, vomiting, headaches, weakness, delirium, hypotension, shock, anemia, and leukopenia	Natural deposits in the earth or from agricultural and industrial practices	Urine, blood	Indiscriminate disposal of arsenic compounds; back siphonage; contaminated groundwater
Cadmium poisoning, acute (see also below CHRONIC ILLNESSES)	Cadmium	15–30 min	Nausea, vomiting, abdominal cramps, diarrhea, shock	Natural deposits such as ores containing other elements, enters water through mining and industry	Blood, urine, hair, or nails	Corrosion of galvanized pipes; erosion of natural deposits (sea salt aerosols); volcanic eruptions; discharge from metal refineries; runoff from waste batteries and paints
Copper poisoning, acute (see also below CHRONIC ILLNESSES)	Copper	A few minutes to a few hours	Acute, gastrointestinal distress	Copper pipes and fittings	Vomitus, gastric washing, urine, blood	Corrosion of the plumbing materials; carbonated beverages; acidic water

Fluoride poisoning (see also below CHRONIC ILLNESSES)	Sodium fluoride	A few minutes to 2 h	Acute, salty or soapy taste, numbness of mouth, vomiting, diarrhea, dilated pupils, spasms, pallor, shock, collapse	Excess addition of sodium fluoride for dental health; natural deposits	Vomitus, gastric washing	Malfunctioning fluoride equipment at water treatment plant; erosion of natural deposits
Incubation(latency) period between 13 and 72 h						
Viruses						
Norovirus infection (Norwalk virus, Norwalk-like virus, Small Round Structured Viruses [SRSV])	Norovirus, Caliciviruses, serotypes, strains and isolates include Norwalk virus; Hawaii virus; Snow Mountain virus; Mexico virus; Desert Shield virus; Southampton virus; Lordsdale virus; Wilkinson virus	Typically 24–48 h	Nausea, vomiting, diarrhea, abdominal pain, myalgia, headache, malaise, low-grade fever, duration up to 60 h	Human feces	Stools, vomitus	Inadequate sewage disposal; using contaminated water
LOWER GASTROINTESTINAL TRACT SIGNS AND SYMPTOMS [ABDOMINAL CRAMPS, DIARRHEA] PREDOMINATE						
Incubation (latency) period usually between 18 and 72 h						
Bacteria						
Aeromonas diarrhea	Aeromonas hydrophila	1–2 days	Watery diarrhea, abdominal pain, nausea, chills, headache	Aquatic environment, both freshwater and marine	Stools	Drinking contaminated, untreated surface water; contamination of foods by sea or surface water

(continued)

Table B (continued)

Illness	Agent	Incubation or latency period[a]	Signs and symptoms[a]	Source of contaminated water	Specimens to collect	Factors contributing to outbreaks
Campylobacteriosis	*Campylobacter jejuni, C. coli*	2–7 days, usually 3–5 days	Abdominal cramps, diarrhea (blood and mucus frequently in stools), malaise, headache, myalgia, fever, anorexia, nausea, vomiting. Sequela: Guillain-Barré syndrome	Mammal, poultry, waterfowl and human feces	Stools, rectal swabs, blood	Using contaminated water supplies (e.g., streams, untreated surface supplies); inadequate disposal of animal feces (also, foodborne)
Cholera	*Vibrio cholerae* serogroup O1 classical and El Tor biotypes; serogroup O139	1–5 days, usually 2–3 days	Profuse, watery diarrhea, (rice-water stools), vomiting, abdominal pain, rapid dehydration, thirst, collapse, reduced skin turgor, wrinkled fingers, sunken eyes, acidosis	Human feces; foods washed or prepared with contaminated water	Stools, rectal swabs	Obtaining fish and shellfish from sewage-contaminated waters in endemic areas; using contaminated water to wash or freshen foods; improper sewage disposal; (also, foodborne)
Cholera-like vibrio gastroenteritis	Non-O1/O139 *V. cholerae* and related spp. (e.g., *V. mimicus, V. fluvialis, V. hollisae*)	1–5 days	Watery diarrhea (varies from loose stools to cholera-like diarrhea)	Human feces, domestic sewage, sea water	Stools, rectal swabs	Inadequate sewage disposal; using contaminated water supply (also, foodborne)

Enterohemorrhagic or Shiga-toxin producing *Escherichia coli* diarrhea	*E. coli* O157:H7, other serotypes non-O157 STEC O26, O45, O103, O104, O111, O113, O121, O128, O145	1–10 days, typically 2–5 days	Watery diarrhea, followed by bloody diarrhea; severe abdominal pain, blood in urine. Sequelae: hemorrhagic colitis, hemolytic uremic syndrome	Cattle, human feces, sewage, cattle manure	Stools, rectal swabs	Inadequate sewage disposal, using contaminated water supply; cattle waste reaching water (also, foodborne)
Enteroinvasive *Escherichia coli* diarrhea	Enteroinvasive *E. coli* strains	½–3 days	Severe abdominal cramps, fever, watery diarrhea (blood and mucus usually present), tenesmus, malaise	Human feces and sewage	Stools, rectal swab	Inadequate sewage disposal; using contaminated water supply (also, foodborne and person-to-person spread)
Enteropathogenic *Escherichia coli* (EPEC), infantile diarrhea	Enteropathogenic *E.coli* strains	Minimum 0.5 h – 34.0 h with a mean 12.9 h (range 4.5–24.0)	Profuse watery diarrhea, sometimes prolonged leading to dehydration, sometimes bloody diarrhea	Human feces and sewage	Stools, rectal swab	

(continued)

Table B (continued)

Illness	Agent	Incubation or latency period[a]	Signs and symptoms[a]	Source of contaminated water	Specimens to collect	Factors contributing to outbreaks
Enterotoxigenic *Escherichia coli* diarrhea	Enterotoxigenic *E. coli* strains	½–3 days	Profuse watery diarrhea (blood and mucus absent), abdominal pain, vomiting, prostration, dehydration, low-grade fever, vomiting in a small percentage of cases	Human feces and domestic sewage	Stools, rectal swabs	Inadequate sewage disposal; using contaminated water supply (also, foodborne and person-to-person spread)
Plesiomonas Enteritis	*Plesiomonas shigelloides*	1–2 days	Diarrhea (blood and mucus in stools), abdominal pain, nausea, chills, fever, headache, vomiting	Aquatic environment, both freshwater and marine	Stools, rectal swabs	Drinking contaminated; untreated surface water; contamination of foods by sea or surface water

Salmonellosis	*Salmonella* (>2000 serovars) from feces of infected animals, environment	6–72 h, typically 18–36 h can be longer than 72 h	Abdominal pain, diarrhea, chills, fever, nausea, vomiting, malaise	Animal and human feces, domestic sewage, farm runoff, meat and poultry processing plant wastes	Stools, rectal swabs	Inadequate sewage disposal; access of animals into well pits and streams, using contaminated water supply (also, foodborne and person-to-person)
Shigellosis	*Shigella dysenteriae, S. flexneri, S. boydii, S. sonnei*	½–7 days, typically 1–3 days	Abdominal pain, diarrhea (stools may contain blood, pus, and mucus), tenesmus, fever, vomiting	Human feces, rectal swabs	Stools, rectal swabs	Inadequate sewage disposal; contaminated well water; cross connections; back siphonage; interrupted disinfection; swimming in polluted waters; freshening produce with contaminated water (also, foodborne and person-to-person)

(continued)

Table B (continued)

Illness	Agent	Incubation or latency period[a]	Signs and symptoms[a]	Source of contaminated water	Specimens to collect	Factors contributing to outbreaks (continued)
Yersiniosis	Yersinia enterocolitica Y. pseudotuberculosis (unknown as a cause of illness from water)	1–7 days	Abdominal pain (may simulate acute appendicitis); low-grade fever, headache, malaise, anorexia, chills, diarrhea, nausea, vomiting	Urine and feces of infected animals (pigs, rodents)	Stools, rectal swabs	Using contaminated surface or spring water; access of animals to surface water (also, foodborne)
Viruses						
Astrovirus gastroenteritis	Astroviruses	1–2 days	Diarrhea, sometimes accomplished by one or more enteric signs or symptoms	Human feces	Stools, acute and convalescent blood	Inadequate sewage disposal; using contaminated water supply (also, foodborne and person-to-person spread)
Enterovirus (see entry under "Generalized infection signs and symptoms")						
Norovirus (see entry under "Upper gastrointestinal signs and symptoms")						
Rotavirus infection	Rotavirus	1–3 days, usually 2 days	Severe watery diarrhea, often with vomiting, fever, abdominal pain, loss of appetite, dehydration	Human feces, domestic sewage	Stools, rectal swabs	Inadequate sewage disposal, using contaminated water supply (also, foodborne and person-to-person spread)

Incubation periods from a few days to a few weeks

Parasites

Amebiasis	*Entamoeba histolytica*	Few days, to several months, typically 2–4 weeks	Mild to severe gastroenteritis, abdominal pain, constipation or diarrhea (stools contain blood and mucus), fever, chills, skin ulcers	Human feces, domestic sewage	Stools, rectal swabs, blood, imaging	Inadequate sewage disposal; using contaminated water supply; cross connections; back siphonage; water and sewer lines in same pits; inadequate disinfection; inadequate filtration of water (also, foodborne and person-to-person spread)
Balantidiasis	*Balantidium coli*		Non-bloody diarrhea, cramping, halitosis, abdominal pain, bloody and/or mucus stools	Human feces, domestic sewage, pig feces, pig manure	Stools	Inadequate sewage disposal; allowing pigs to access areas where crops for human consumption are grown; contact with water frequented by pigs
Cryptosporidiosis	*Cryptosporidium* spp.	1–12 days, typically 8 days	Prolonged watery diarrhea, abdominal pain, anorexia, vomiting, low-grade fever	Human, cattle, deer, sheep, cat, rodent, and other animal feces, domestic sewage, animal waste	Stools, intestinal biopsy	Inadequate sewage or animal waste disposal; contaminated water; inadequate filtration of water

(continued)

Table B (continued)

Illness	Agent	Incubation or latency period[a]	Signs and symptoms[a]	Source of contaminated water	Specimens to collect	Factors contributing to outbreaks
Cyclosporiasis	*Cyclospora cayetanensis*	1–11 days, typically 7 days	Prolonged watery diarrhea, weight loss, fatigue, nausea, anorexia, abdominal cramps	Human feces, domestic sewage	Stools	Sewage contaminated irrigation water or spraying water suspected; washing fruits with contaminated water; untreated water
Giardiasis	*Giardia intestinalis* (also known as *Giardia lamblia*, *Giardia duodenalis*)	5–25 days, typically 7–10 days	Diarrhea (pale, greasy, malodorous stools), abdominal pain, bloating, nausea, weakness, vomiting, dehydration, fatigue, weight loss, fever	Human feces, domestic sewage, animal (e.g., beaver) feces	Stools	Inadequate sewage disposal, using untreated surface water supplies (e.g., mountain streams and lakes); inadequate or interrupted disinfection; cross connections; improper filtration; using untreated surface water supplies as ingredient for processing (also, foodborne and person-to-person)

SIGNS OF SUFFOCATION (CYANOSIS) OCCUR

Variable incubation periods depending on concentration ingested

Chemicals

Methemoglobinemia (Blue baby syndrome)	Nitrites and nitrates	Variable depending on nitrate concentration in water and amount ingested	Bluish coloration of skin, brownish color of blood (usually occurs in infants less than 4 months of age)	Fertilizers, domestic sewage, animal feces	Blood	Shallow, unprotected, or cased wells; excessive use of fertilizers in water shed; large numbers of animals around well

NEUROLOGICAL SYMPTOMS AND SIGNS (VISUAL DISTURBANCES, TINGLING, AND/OR PARALYSIS OCCUR)

Incubation (latency) period usually less than 1 h

Chemicals

Organophosphate poisoning	Organic phosphorous insecticides	Few minutes to a few hours	Nausea, vomiting, abdominal cramps, diarrhea, headache, nervousness, blurred vision, chest pain, cyanosis, confusion, twitching, convulsions	Pesticides	Blood	Back siphonage of insecticide compounds from hose application; seepage after soil-foundation spraying (also foodborne under similar situations of contamination)

(continued)

Table B (continued)

Illness	Agent	Incubation or latency period[a]	Signs and symptoms[a]	Source of contaminated water	Specimens to collect	Factors contributing to outbreaks
Incubation (latency) period usually between 1 and 6 h						
Chemicals						
Chlorinated hydrocarbon poisoning	Chlorinated hydrocarbon insecticides	30 min to 6 h	Nausea, vomiting, paresthesia, dizziness, muscular weakness, anorexia, weight loss, confusion	Pesticides	Blood, urine, stools, gastric washings	Back siphonage of insecticide compounds from hose application; seepage after soil-foundation spraying (also foodborne under similar situations of contamination)
SKIN INVOLVEMENT						
Incubation period greater than 1 week						
Parasites						
Dracunculiasis	*Dracunculus medinensis*	8–14 months, usually 12 months	Blister, burning and itching at site of exit, especially the foot (vesicle and milky fluid), fever, nausea, vomiting, diarrhea, dyspnea, eosinophilia	Larvae discharge from worms protruding from skin of infected person	Visualization of adult worm under skin	*Cyclops* (crustacean) in water supply; unprotected water supply; frequently step wells and ponds

Incubation period usually less than 1 week

Bacteria

Tularemia	*Francisella tularensis*	1–10 days, usually 3 days	Sore throat, mouth ulcers, tonsillitis, and swelling of lymph glands in the neck (from ingestion of contaminated water)	Blood and tissue of infected wild mammals	Blood	Access of animals to water supplies; using contaminated stream or untreated surface water; dead animals in surface water

Incubation period usually greater than 1 week

Bacteria

Typhoid and paratyphoid fevers	*Salmonella* Typhi for typhoid; *S.* Paratyphi A, B, and C for paratyphoid fever	7–28 days, usually 14 days	Continued fever, malaise, headache, cough, nausea, vomiting, anorexia, abdominal pain, chills, rose spots, constipation or bloody diarrhea. Sequela: reactive arthritis	Human feces and urine, domestic sewage	Stools, rectal swabs, blood in incubatory and early acute phase, urine in acute phase	Inadequate sewage disposal; back siphonage; cross contamination; using contaminated water supplies; chlorination failures (also, foodborne and person-to-person)

(continued)

Table B (continued)

Illness	Agent	Incubation or latency period[a]	Signs and symptoms[a]	Source of contaminated water	Specimens to collect	Factors contributing to outbreaks
Viruses						
Enterovirus infection	Poliovirus Coxsackieviruses Echoviruses	3–14 days, usually 5–10 days	Variable, including fever, respiratory tract symptoms, meningitis, herpangina, pleurodynia, conjunctivitis, myocardiopathy, diarrhea, paralysis, encephalitis, ataxia	Human feces and domestic sewage, respiratory discharges	Stools, rectal swabs, respiratory secretions, blood	Fecal contamination of water
Hepatitis A	Hepatitis A virus	15–50 days, usually 25–30 days	Fever, malaise, lassitude, anorexia, nausea, abdominal pain, jaundice, dark urine, light-colored stools	Human feces and urine, domestic sewage	Stools, urine, blood	Inadequate sewage disposal; using contaminated surface or groundwater supplies; cross connections, back siphonage, inadequate disinfection; harvesting shellfish from sewage contaminated water (also, foodborne and person-to-person)

Hepatitis E	Hepatitis E virus	15–65 days, usually 35–40 days	Similar to above (high mortality for pregnant women in third trimester due to fulminant hepatitis)	Human feces and domestic sewage	Stools, blood	Inadequate sewage disposal; using contaminated surface water or groundwater supplies; cross connections, back siphonage, inadequate disinfection; harvesting shellfish from sewage contaminated water (also, foodborne and person-to-person)
Parasites						
Toxoplasmosis	*Toxoplasma gondii*	4–28 days, mean 9 days	Fever, headache, myalgia, rash	Cat (family) feces	Blood	Surface water not filtered

(continued)

Table B (continued)

Illness	Agent	Incubation or latency period[a]	Signs and symptoms[a]	Source of contaminated water	Specimens to collect	Factors contributing to outbreaks
ACUTE AND/OR CHRONIC ILLNESS Variable incubation periods *Algae*						
Cyanobacterial toxin poisoning (blue-green algae producing microcystins)	Some strains of the following cyanobacteria: Hepatotoxins—*Anabaena* spp., *Microcystis* spp. *Oscillatoria* spp., *Nodularia* spp., *Nostoc* spp., *Cylindrospermopsis* spp. and *Umezakia* spp., Neurotoxins—*Aphanizomenon* spp., *Oscillatoria* spp., Toxic alkaloids—*Cylindrospermopsis raciborskii*	Few hours, but can be weeks months, or years	Can be in stages: nausea, vomiting, fever, headache; followed by a pulmonary stage, then hepatotoxicosis and multiple organ failure. Other symptoms include muscle and joint pain, blisters, mouth ulcers, and allergic reactions, and seizures	Cyanobacterial blooms in water	Blood	Ingestion of water containing cyanobacterial toxins; water plant to change water source (if possible); adjust water intake depth; activated carbon to remove toxins
CHRONIC ILLNESS *Chemicals*						
Arsenic poisoning (see also above UPPER GASTROINTESTINAL SIGNS AND SYMPTOMS)	Arsenic (inorganic)		Skin changes, links to cancer of the bladder, lungs, skin, kidneys, nasal passages, liver and prostate (only some of signs listed here)	Natural deposits in the earth or from agricultural and industrial practices	Urine, blood	Contaminated groundwater

Cadmium poisoning (see also above UPPER GASTROINTESTINAL SIGNS AND SYMPTOMS)	Cadmium	Several months	Kidney damage, lung damage, fragile bones	Natural deposits such as ores containing other elements, enters water through mining and industry	Blood, urine, hair, or nails	Corrosion of galvanized pipes; erosion of natural deposits (sea salt aerosols); volcanic eruptions; discharge from metal refineries; runoff from waste batteries and paints
Copper poisoning (see also above UPPER GASTROINTESTINAL SIGNS AND SYMPTOMS)	Copper		Liver or kidney damage	Natural deposits such as ores containing other elements	Vomitus, gastric washing, urine, blood	Corrosion of the plumbing materials
Fluoride poisoning (see also above UPPER GASTROINTESTINAL SIGNS AND SYMPTOMS)	Sodium fluoride	Months to years	Bone disease (including pain and tenderness of the bones); children may get mottled teeth	Excess addition of sodium fluoride for dental health	Vomitus, gastric washing	Malfunctioning fluoride equipment at water treatment plant; erosion of natural deposits
Lead poisoning	Lead	Weeks to months	Infants and children: delays in physical or mental development; children could show slight deficits in attention span and learning abilities; Adults: kidney problems, high blood pressure	Household plumbing materials in water service lines	Blood	Plumbing materials made of lead or with lead solder; potable water from a corrosive source without treatment reacting with lead pipes

[a]This list is not exhaustive. Not all of the listed illnesses will occur in all countries. Cultural, economic, and climatic conditions preclude or foster conditions that influence their occurrence

Table C Illnesses acquired by contact with water: a condensed classification by, symptoms, incubation period, and types of agents

Illness	Etiologic agent	Incubation or latency period	Signs and Symptoms	Source of contaminated water	Specimens to collect	Factors contributing to waterborne outbreaks
						Swimming in fresh water, breaks in skin
SKIN INFECTIONS/CONDITIONS (RASHES, BLISTERS, AND/OR PUSTULES) OCCUR						
Incubation period of a few hours to a few days						
Bacteria						
	Aeromonas hydrophila	Few hours to a few days	Cellulits, septicemia	Fresh and brackish water	Exudate from skin lesions	
Pseudomonas aeruginosa infection (hot tub rash, Pseudomonas dermatitis, Pseudomonas folliculitis) (swimmer's ear, otitis externa)	Pseudomonas aeruginosa	Few hours to a few days	Generalized rash, pustules, most prevalent in areas covered by bathing suit, and in ears, earache	Water, skin	Exudate from skin lesion or ear	Inadequate cleaning of pools, water vessels, and devices; immersion in improperly cleaned and disinfected whirlpool baths or hot tubs
Staphylococcus infection (hot tub rash) (swimmer's ear, otitis externa)	Staphylococcus spp.	Within a few days of swimming	Rash, and in ears, earache	Water, skin	Exudate from skin lesion or ear	Inadequate cleaning of pools, water vessels, and devices; immersion in improperly cleaned and disinfected whirlpool baths or hot tubs
Tularemia	Francisella tularensis	1–10 days, usually 3 days	Ulcers on skin, chills, high fever	Blood and tissue of infected wild mammals, or arthropods	Blood, sputum, lymph node biopsy, muscle tissue	Access of animals to water courses, dead animals in surface water, contact with contaminated water
Vibriosis	Vibrio alginolyticus, Vibrio parahaemolyticus, Vibrio mimicus	Few hours to few days	Wound and ear infections	Sea water	Exudates from skin lesions	Swimming in sea water, breaks in skin, warm weather

Vibrio vulnificus infections	*Vibrio vulnificus*	Average 16 h	Wound infections, septicemia; fever, chills, malaise, prostration; metastatic cutaneous lesions (preexisting liver disease is often associated with illness)	Sea water	Exudates from skin lesions, blood	Swimming in sea water, breaks in skin, chronic liver illness with declining immune defense mechanisms
Parasites						
Amebic abscesses	*Acanthamoeba* spp. *Balamuthia* spp.	Unknown	Painful subcutaneous abscesses	Water	Swabs or aspirates of lesions	Exposure to warm muddy waters or industrial water wash solutions, patients are often immunocompromised
Schistosoma dermatitis (swimmer's itch)	*Schistosoma* cercaria, many spp.	Few minutes to hours	Dermatitis, prickly sensation and intense itching, redness, blisters	Feces and urine of infected animals and birds	None	Snails in water, swimming and wading in infested waters
Algae						
Cyanobacterial toxin poisoning	Cyanobacteria	Several days or more	Rash, hives, skin blisters (especially on the lips and under swimsuits), mouth ulcers, eye/ear irritation	Algal blooms	None	Skin exposure during bathing and swimming to water where algae has bloomed

(continued)

Table C (continued)

Illness	Etiologic agent	Incubation or latency period	Signs and Symptoms	Source of contaminated water	Specimens to collect	Factors contributing to waterborne outbreaks
Seaweed dermatitis (swimmer's itch)	*Lyngbya majuscula*	Few minutes to hours	Redness, blisters and lesions containing pus on skin	Sea water natural habitat	None	Warm weather, swimming in sea water with blue-green algae bloom

EYE INFECTIONS (CONJUNCTIVITIS, PURULENT DISCHARGES, REDNESS, SWELLING OCCUR)
Incubation period usually less than a week

Bacteria						
Conjunctivitis	*Chlamydia trachomatis*	5–12 days	Conjunctivitis, purulent discharge	Genitourinary exudates	Skin or conjunctival swabs	Unchlorinated swimming pool water
Pseudomonas aeruginosa infection	*Pseudomonas aeruginosa*	Few hours to a few days	Conjunctivitis	Water, skin	Corneal scraping	Rubbing eyes while immersed in hot tubs, contaminated lens solution
Viruses						
Enteroviral hemorrhagic conjunctivitis	Adenoviruses, Picornaviruses	4–12 days ½–3 days	Redness, swelling, pain in eye; hemorrhages on conjunctiva	Discharges from infected eyes	Exudates from eye, blood	Unchlorinated swimming pool water
Parasites						
Amebic keratitis	*Acanthamoeba* spp.	Unknown – probably weeks to months	Keratitis, severe eye pain, eye redness, blurred vision, sensitivity to light	Shallow muddy ponds	Corneal scrapings	Wearing contact lenses while swimming in fresh water, using non-sterile saline (homemade) for contact lenses
Chemicals						
Swimming pool conjunctivitis	Chemical added to cause maladjusted pH	Few minutes	Redness of eyes, swelling around eyes	Water and chemicals used for treatment of pools		Improper pH adjustment of swimming pool water

GENERALIZED INFECTION SIGNS AND SYMPTOMS (FEVER, CHILLS AND/OR MALAISE OCCUR)

Incubation period less than 1 week
Parasites

Primary amebic meningoencephalitis	*Naegleria fowleri, Balamuthia mandrillaris*	1–7 days	Severe frontal headache, nausea, high fever, meningitis, death	Warm fresh bodies of water, hot springs	Spinal fluid, biopsy	Recent swimming in fresh water during hot water, sniffing fresh water up nose

Incubation period greater than 1 week
Parasites

Granulomatous amebic encephalitis (GAE)	*Acanthamoeba* spp. *Balamuthia* spp.	Weeks to months	Parasites invade through skin or lungs and then CNS. Headaches, seizures, stiff neck. Lesions may form in internal organs	Water		Affects chronically ill or debilitated

Bacteria

Chromobacterium violaceum infection	*Chromobacterium violaceum*	3–14 days	Fever, malaise, myalgia, abdominal pain, pustular rash, sepsis with hepatic abscess	Soil, water	Blood	Breaks in skin, puncture wound

(continued)

Table C (continued)

Illness	Etiologic agent	Incubation or latency period	Signs and Symptoms	Source of contaminated water	Specimens to collect	Factors contributing to waterborne outbreaks
Granulomatous tuberculous lesions	*Mycobacterium marinum*	3–6 weeks	Small papular lesion of spongy consistency with crusty scale surrounding it. Thick secretions under crust, leaving bluish red, soft scar at elbow, knees, and ankles, foot, hand, bridge of nose	Soil, water	Exudate from skin lesions	Rough walls of swimming pools, inadequate chlorination of pool water, injury, scrape, puncture during swimming; abrasions when cleaning aquariums, immersion in natural bathing waters (fresh and salt)
Leptospirosis (Weil's disease)	*Leptospira* spp.	4–19 days, usually 10 days	Fever, chills, malaise, muscular aches, headache, vomiting, stiff neck, occasionally a rash	Animal (e.g., dog, rodent, cattle, deer, swine, fox, squirrel, skunk, raccoon, opossum, muskrat urine)	Urine, blood	Wound and abrasion and mucus membrane entrance. Wading or swimming in farm ponds or water to which animals have access

Parasites

Schistosomiasis (Bilharziasis)	Schistosoma haemotobium	4–6 weeks	Blood in urine, fever, hepatic pain, epigastric distress, diarrhea, bladder obstruction	Urine of infected human	Urine	Inadequate sewage control, snails in water, water pollution, swimming or bathing in human sewage contaminated water
	S. japonicum S. mansoni S. intercalatum S. mekongi	4–6 weeks	Fever, chills, night sweats, enlarged tender liver, pain in back, groin, legs, diarrhea, high eosinophilic count, epigastric distress	Feces of infected person; feces of pigs, cattle, buffalo, dogs, rodents, cats, horses	Stool, urine	Inadequate sewage control, access of animals to water courses, snails in water, water pollution, swimming, wading, or bathing in sewage contaminated water

Table D Illnesses acquired by inhalation of microorganisms aerosolized from water[a]. A classification by symptoms, incubation period and type of agent

Illness and classification	Agent	Incubation or latency period	Signs and symptoms	Source of contaminated water	Specimens to collect	Factors contributing to outbreaks
Legionellosis, Legionnaires' Disease	*Legionella pneumophila*, other *Legionella* spp.	2–10 days	Muscle pain, loss of appetite, headache, high fever, dry cough, chills, confusion, disorientation, nausea, diarrhea, and vomiting, chest pain difficulty breathing	Water, including groundwater, fresh and marine surface waters, potable (treated) water, cooling towers, evaporative condensers and whirlpools	Respiratory secretions, lung biopsy, urine, abnormal X-ray, blood serology	Breathing aerosol produced from water containing systems; lack of maintenance of water systems; dead ends in water systems; lack of disinfection in water systems; lack of cleaning biofilm from water system surfaces
Pontiac Fever	*Legionella* spp.	24–48 h	Many of the above symptoms, but less severe, no pneumonia or death	Water, including groundwater, fresh and marine surface waters, potable (treated) water, cooling towers, evaporative condensers and whirlpools	Urine, blood serology	Breathing aerosol produced from water containing systems such as jacuzzis, decorative fountains, cooling towers; cleaning in a confined space with high pressure water; taking a shower

[a]Refers to infections caused by airborne bacteria that are amplified in water

Table E Guidelines for Specimen Collection[a]

Instructions for collecting stool specimens[b]

Instructions	Bacterial	Parasitic[c]	Viral[d]	Chemical
When to collect	During period of active diarrhea (preferably as soon as possible after onset of illness)	Anytime after onset of illness (preferably as soon as possible)	Within 48–72 h after onset of illness	Soon after onset of illness (preferably within 48 h of exposure to contaminant)
How much to collect	Two rectal swabs or swabs of fresh stool from ten ill persons; samples from ten controls also can be submitted. Whole stool is preferred if nonbacterial stool testing considered	A fresh stool sample from ten ill persons; samples from ten controls can also be submitted. To enhanced detection, three stool specimens per patient can be collected >48 h apart	As much stool sample as possible from ten ill persons; samples from ten controls also can be obtained from ten controls. A minimum of 10 mL of stool sample from each; samples also	A fresh urine sample (50 mL) from ten ill persons; samples from ten controls also can be submitted. Collect vomitus, if vomiting occurs within 12 h of exposure. Collect 5–10 mL whole blood if a toxin/poison is suspected that is not excreted in urine
Method for collection	For rectal swabs, moisten two swabs in an appropriate transport medium (e.g., Cary-Blair, Stuart, Amies; buffered glycerol-saline is suitable for *E. coli, Salmonella, Shigella,* and *Y. enterocolitica* (but not for *Campylobacter* and *Vibrio*). Insert swab 2.5–4 cm (1–1.5 in) into rectum and gently rotate. Place both swabs into the same tube deep enough that medium covers the cotton tips. Break off top portion of sticks and discard. Alternatively, swab whole stools and put them into Cary-Blair medium	Collect bulk stool specimens, unmixed with urine, in a clean container. Place a portion of each stool sample into 10 % formalin and polyvinyl alcohol preservative (PVA) at a ratio of one part stool to three parts preservative. Mix well. Save portion of the unpreserved stool placed into leak proof container for antigen or PCR testing	Place fresh stool specimens (liquid preferable), unmixed with urine, into clean, dry containers, e.g., urine specimen cups	Collect urine, blood, or vomitus in prescreened containers[e]. If prescreened containers are not available, submit field blanks with samples[f]. Most analyses from blood requires separation of serum from red cells. Cyanide, lead and mercury analyses require whole blood collected in prescreened EDTA tubes. Volatile organic compounds require whole blood collected in a specially prepared gray-top tube

(continued)

Table E (continued)

Instructions	Bacterial	Parasitic[c]	Viral[d]	Chemical
Instructions for collecting stool specimens[b]				
Storage of specimens after collection	Refrigerate swabs in transport media at 4 °C. When possible test within 48 h after collection; otherwise, freeze samples at −70 °C. Refrigerate whole stool, process it within 2 h after collection. Store portion of each stool specimen frozen at less than −15 °C for antigen or PCR testing	Store specimen in fixative at room temperature, or refrigerate unpreserved specimen at 4 °C. A portion of unpreserved stool specimen may be frozen at less than−15 °C for antigen or PCR testing	Immediately refrigerate at 4 °C. Store portion of each stool specimen frozen at less than −15 °C for antigen or PCR testing	Immediately refrigerate at 4 °C and if possible freeze urine, serum, and vomitus specimens at less than −15°C. Refrigerate whole blood for volatile organic compounds and metals at 4 °C
Transportation	For refrigeration: Follow instructions for viral samples. For frozen samples: Place bagged and sealed samples on dry ice. Mail in insulated box by overnight mail	For refrigeration: Follow instructions for viral samples. For room-temperature samples: Mail in waterproof container	Keep refrigerated. Place bagged and sealed specimens on ice or with frozen refrigerant packs in an insulated box. Send by overnight mail. Send frozen specimens on dry ice for antigen or PCR testing	Immediately refrigerate at 4 °C and if possible freeze urine, serum, and vomitus specimens at less than −15 °C. Refrigerate whole blood for volatile organic compounds and metals at 4 °C. Place double bagged and sealed urine, serum, and vomitus specimens on dry ice. Mail in an insulated box by overnight mail. Ship whole blood in an insulated container with prefrozen ice packs. Avoid placing specimens directly on ice packs

[a]From CDC webpage

[b]Label each specimen in a waterproof manner and put the samples in sealed, waterproof containers (i.e., plastic bags). Batch the collection and send in overnight mail to arrive at the testing laboratory on a weekday during business hours unless other arrangements have been made in advance with the testing laboratory. Contact the testing laboratory before shipping, and give the testing laboratory as much advance notice as possible so that testing can begin as soon as samples arrive. When etiology is unclear and syndrome is nonspecific, all four types of specimens may be appropriate to collect

[c]For more detailed instructions on how to collect specimens for specific parasites, please go to http://www.CDC.gov and search the website of key words

[d]For more detailed instructions on how to collect specimens for viral testing, please go to http://www.CDC.gov and search the website of key words

[e]The containers have been tested for the presence of the chemical of interest prior to use

[f]Unused specimen collection containers that have been brought in to the field and subjected to the same field conditions as the used containers. These containers are then tested for trace amounts of the chemical of interest

Table F General instructions for collecting water samples for microbiological analysis

	Type of agent to be tested for		
	Viruses	Bacteria	Parasites
When to collect	As soon as hypothesis of waterborne outbreak formulated	As soon as hypothesis of waterborne outbreak formulated	As soon as hypothesis of waterborne outbreak formulated
How much to collect	400 L (surface water up to 360 L; groundwater up to 1500–1800 L)	1 L per pathogen that is to be sought and an additional 200 mL for testing for indicator organisms. Collect at least three samples from each well and 8–10 samples from distribution system	400 L
Method of collection	Pump through electropositive cartridge filters (approximately 10 L per minute rate do not surpass manufacturers rated flow rate for filter type)	Collect in sample bottles or bags (see test for procedures)	Pump through yarn-wound cartridge filters
Storage of sample after collection	Immediately refrigerate and hold at 4°C. Process within 72 h. Do not freeze	Immediately refrigerate and hold at 4°C. Testing should be done within 24 h after collection	Refrigerate at 4°C
Transportation	Keep at 4°C—use frozen refrigerant packs in an insulated box	Keep at 4°C—use frozen refrigerant packs in an insulated box. For frozen samples, put samples on dry ice in insulated box. Either bring to laboratory day of collection or send by courier	Keep at 4°C—use frozen refrigerant packs in an insulated box

Table G Guidelines for confirmation of waterborne outbreaks[a]

Bacterial/chemical/parasitic/viral

Etiologic agent	Incubation period	Clinical syndrome	Confirmation
Bacterial			
1. *Campylobacter jejuni/coli*	2–10 days; usually 2–5 days	Diarrhea (often bloody), abdominal pain, fever	Isolation of organism from clinical specimens from two or more ill persons OR Isolation of organism from epidemiologically implicated water
2. *Escherichia coli* (a) Enterohemorrhagic (*E. coli* O157:H7 and others)	1–10 days; usually 3–4 days	Diarrhea (often bloody), abdominal cramps (often severe), little or no fever	Isolation of *E. coli* O157:H7 or other Shiga-like toxin (verocytotoxin) producing *E. coli* from clinical specimens from two or more ill persons OR Isolation of *E. coli* O157:H7 or other Shiga-like toxin (verocytotoxin) from epidemiologically implicated water
(b) Enteroinvasive (EIEC)	Variable	Diarrhea (might be bloody), fever, abdominal cramps	Isolation of same enteroinvasive serotype from stool of two or more ill persons
(c) Enterotoxigenic (ETEC)	6–48 h	Diarrhea, abdominal cramps, nausea, vomiting and fever less common	Isolation of organism of same serotype demonstrated to produce heat stable (ST) and/or heat-labile (LT) enterotoxin from stool of two or more ill persons
(d) Enteropathogenic (EPEC)	Variable	Diarrhea, fever, abdominal cramps	Isolation of organism of same enteropathogenic serotype from stool of two or more ill persons
3. Nontyphoidal *Salmonella*	6 h–10 days; usually 6–48 h	Diarrhea, often with fever and abdominal cramps	Isolation of organism of same serotype from clinical specimens from two or more ill persons OR Isolation of organisms from epidemiologically implicated water
4. *Salmonella* Typhi	3–60 days; usually 7–14 days	Fever, anorexia, malaise, headache and myalgia; sometimes diarrhea or constipation	Isolation of organism from clinical specimens from two or more ill persons OR Isolation of organism from epidemiologically implicated water

5. *Shigella* spp.	12 h–6 days; usually 2–4 days	Diarrhea (often bloody), often accompanied by fever and abdominal cramps	Isolation of organism of same serotype from clinical specimens from two or more ill persons OR Isolation of organism from epidemiologically implicated water
6. *Vibrio cholerae* (a) O1 or O139	1–5 days	Watery diarrhea, often accompanied by vomiting	Isolation of toxigenic organism from stool or vomitus of two or more persons OR Significant rise in vibriocidal, bacterial-agglutinating, or antitoxin antibodies in acute- and early convalescent-phase sera among persons not recently immunized OR Isolation of toxigenic organism from epidemiologically implicated water
(b) Non-O1 and non-O139	1–5 days	Watery diarrhea	Isolation of organism of same serotype from stool of two or more ill persons
7. *Yersinia enterocolitica*	1–10 days; usually 4–6 days	Gastrointestinal symptoms, abdominal pain (often severe) mimicking appendicitis), diarrhea	Isolation of organism from clinical specimens from two or more persons OR Isolation of pathogenic strain of organism from epidemiologically implicated water
Chemical 1. Heavy metals • Antimony • Arsenic • Cadmium • Copper • Fluoride	5 min–8 h; usually <1 h	Vomiting, often metallic taste	Demonstration of high concentration of metal in epidemiologically linked water

(continued)

Table G (continued)

Bacterial/chemical/parasitic/viral

Etiologic agent	Incubation period	Clinical syndrome	Confirmation
Parasitic			
1. *Cryptosporidium* spp.	2–28 days; median: 7 days	Diarrhea, nausea, vomiting, fever	Demonstration of oocysts in stool or in small-bowel biopsy of two or more ill persons OR Demonstration of organism from epidemiologically implicated water
2. *Cyclospora cayetanensis*	1–14 days; median: 7 days	Diarrhea, nausea, anorexia, weight loss, cramps, gas, fatigue, low-grade fever; may be relapsing or protracted	Demonstration of the parasite by microscopy or molecular methods in stool or in intestinal aspirate or biopsy specimens from two or more ill persons OR Demonstration of the parasite from epidemiologically implicated water
3. *Giardia intestinalis*	3–25 days; median: 7 days	Diarrhea, gas, cramps, nausea, fatigue	Demonstration of the parasite in stool or small-bowel biopsy specimen of two or more ill persons
Viral			
1. Astrovirus	12–48 h	Diarrhea, vomiting, nausea, abdominal cramps, low-grade fever	Detection of viral RNA in at least two bulk stool or vomitus specimens by real-time or conventional reverse transcriptase-polymerase chain reaction (RT-PCR) OR Visualization of viruses with characteristic morphology by electron microscopy in at least two or more bulk stool or vomitus specimens OR Two or more stools positive by commercial enzyme immunoassay (EIA)

| 2. Hepatitis A | 15–50 days; median: 28 days | Jaundice, dark urine, fatigue, anorexia, nausea | Detection of immunoglobulin M antibody to hepatitis A virus (IgM anti-HAV) in serum from two or more persons who consumed epidemiologically implicated water |
| 3. Norovirus (NoV) | 12–48 h; median: 33 h | Diarrhea, vomiting, nausea, abdominal cramps, low-grade fever | Detection of viral RNA in at least two bulk stool or vomitus specimens by real-time or conventional reverse transcriptase-polymerase chain reaction (RT-PCR) OR Visualization of viruses (NoV) with characteristic morphology by electron microscopy in at least two or more bulk stool or vomitus specimens OR Two or more stools positive by commercial enzyme immunoassay (EIA) |

[a]Most etiologic agent descriptions based on information from the Centers for Disease Control and Prevention, Atlanta, GA, USA

Table H Guidelines for confirmation of water responsible for illness

Confirmation status	Criteria
Confirmed vehicle	Isolation of agent from ill persons and from water and laboratory criteria for confirming etiologic agent as stated in Table G. Combination of on-site investigation, statistical evidence and laboratory analysis. (see entries below)
Presumptive vehicle	On-site investigation demonstrating source and mode of contamination of water and survival of etiologic agent in water. Also, desirable to have laboratory isolations from water of etiologic agent that causes syndrome similar to that observed during the investigation and other supportive epidemiologic data. If so, this might provide sufficient evidence for confirmation. OR p-value for water <0.05 when other epidemiologic data supports water hypothesis. Also, desirable to have either laboratory isolations from water or on-site investigation that demonstrates source and mode of contamination and survival of treatment that supports the hypothesis. If so, this might provide sufficient evidence for confirmation. OR Odds ratio or relative risk for water greater than 2 and the lower limit of the 95% confidence level greater than 1 when other epidemiologic data supports the water vehicle hypothesis. Also, desirable to have either laboratory isolations from water or on-site investigation that demonstrates source and mode of contamination and survival of treatment that supports the hypothesis. If so, this might provide sufficient evidence for confirmation.

Table I $CT_{99.9}$(3-log) values for inactivation of *Giardia* cysts at different concentrations of disinfectants, temperatures and pH values[a]

Concentration (mg/L)		Disinfectant						
		Free chlorine				ClO$_2$	Ozone	Chloramine[b]
<0.5°C	pH	6	7	8	9	6–9	6–9	6–9
≤0.4		137	195	277	390			
0.6		141	200	286	407			
1		148	210	304	437			
2		165	236	346	500			
3		181	261	382	552	63	2.9	3800
5°C	pH	6	7	8	9	6–9	6–9	6–9
≤0.4		97	139	198	279			
0.6		100	143	204	291			
1		105	149	216	312			
2		116	165	243	353			
3		126	182	268	389	26	1.9	2200
10°C	pH	6	7	8	9	6–9	6–9	6-9
≤0.4		73	104	149	209			
0.6		75	107	153	218			
1		79	112	162	234			
2		87	124	182	265			
3		95	137	201	292	23	1.43	1850
15°C	pH	6	7	8	9	6–9	6–9	6-9
≤0.4		49	70	99	140			
0.6		50	72	102	146			
1		53	75	108	156			
2		58	83	122	177			
3		63	91	134	195	19	0.95	1500
20°C	pH	6	7	8	9	6–9	6–9	6–9
≤0.4		36	52	74	105			
0.6		38	54	77	109			
1		39	56	81	117			
2		44	62	91	132			
3		47	68	101	146	15	0.72	1100
25°C	pH	6	7	8	9	6–9	6–9	6-9
≤0.4		24	35	50	70			
0.6		25	36	51	73			
1		26	37	53	75			
2		29	41	61	89			
3		32	46	67	97	11	0.48	750

[a]Source: Environmental Protection Agency Technical Guidance Manual LT1ESWTR Disinfection Profiling and Benchmarking. March 2003
[b]Chloramines refer to all forms of chloramine. The *CT* values may be assumed to achieve greater than 99.99% inactivation of viruses only if chlorine is added and mixed in the water before addition of ammonia. If this condition is not met, the system must demonstrate by on-site studies or other information that it is achieving at least this much inactivation of viruses

Table J *CT* values for 99.99% inactivation of viruses at pH 6–9 at different temperatures with different disinfectants[a]

Disinfectant	Temperature (°C)						
	Log inactivation	0.5	5	10	15	20	25
Free chlorine	2	6	4	3	2	1	1
	3	9	6	4	3	2	1
	4	12	8	6	4	3	2
Ozone	2	0.9	0.6	0.5	0.3	0.25	0.15
	3	1.4	0.9	0.8	0.5	0.4	0.25
	4	1.8	1.2	1.0	0.6	0.5	0.3
Chlorine dioxide	2	8.4	5.6	4.2	2.8	2.1	1.4
	3	25.6	17.1	12.8	8.6	6.4	4.3
	4	50.1	33.4	25.1	16.7	12.5	8.4
Chloramines	2	1243	857	643	428	321	214
	3	2063	1423	1067	712	534	356
	4	2883	1988	1491	994	746	497

[a]Source: Environmental Protection Agency Guidance Manual for the Compliance with Filtration and Disinfection Requirements Public Water Systems Using Surface Water Sources. March 1991

Table K Estimated log removal of *Giardia* and viruses by various methods of filtration[a]

Method of filtration	Estimated log removal	
	Giardia (3-log inactivation is goal)	Viruses (4-log inactivation is goal)
Conventional (provided turbidity <0.5 NTU)	2.5	2.0
Direct	2.0	1.0
Slow sand	2.0	2.0
Diatomaceous earth	2.0	1.0

[a]Environmental Protection Agency (Federal Register, June 29, 1989, 40 CFR, Parts 141 and 142) http://water.epa.gov/lawsregs/rulesregs/sdwa/swtr/upload/SWTR.pdf. Accessed May 22, 2015

FOODBORNE, WATERBORNE, ENTERIC ILLNESS COMPLAINT REPORT
Form A

		Complaint no.*

Complaint received from	Address	Phone Home Work
Person to contact for more information	Address	Phone Home Work e-mail

Complaint

Type of complaint:* ☐ Illness ☐ Contaminated/spoiled/adulterated food ☐ Poor quality drinking water
☐ Poor quality recreational water ☐ Unsanitary establishment ☐ Complaint related to media publicity
☐ Disaster ☐ Other (specify)

Illness: ☐ Yes,[1,2*] ☐ No Number ill* _____ Number exposed _____ Time first symptom: Date* _____
Hour _____

Predominant symptoms:* ☐ Vomiting ☐ Diarrhea ☐ Fever ☐ Neurological ☐ Skin ☐ Other (specify)

Physician consulted: ☐ Yes ☐ No If yes, Name	Address	Phone

Hospitalized: ☐ Yes ☐ No Emergency Room visit: ☐ Yes ☐ No
If yes,
Hospital name _____ Address _____
_____ Phone _____
Physician's name _____ Phone _____
Laboratory examination of specimen: Type specimen Organism/Toxin detected*

Suspect food/water* _____ Source of food/water † _____
Brand identification † Code/Lot no. †

Suspect meal, event or place:* _____ Date _____ Time _____
Address Phone

NAME	STATUS	ADDRESS	PHONE
1.	☐ ill ☐ well		
2.	☐ ill ☐ well		
3.	☐ ill ☐ well		
4.	☐ ill ☐ well		

Domestic water source: ☐ Community ☐ Non-community ☐ Bottled water
☐ Stream/lake ☐ Vended ☐ Well ☐ Untreated ☐ Other (specify)

Places and locations where foods eaten past 72 hours, other than home *[3]	Place and locations where water ingested past 2 weeks, other than home *[3]	Place and locations where recreation water contacted past 2 weeks *[3]

History of exposures within past six weeks:* ☐ Domestic travel (Place) _____
☐ International travel (Place) _____ ☐ Child care ☐ Contact with ill person outside
household or ill person visited household (indicate name) ☐ Contact with ill person within household
(indicate name) ☐ Ill animal _____

Received by	Date of complaint/alert	Time	Disposition

Investigator's name	Comments

[1] If yes, public health professional staff member should obtain information about patient which should be put on Form C.
[2] Ask person to collect vomitus and/or stool in a clean jar, wrap, identify, and refrigerate; hold until health official makes further arrangements.
[3] Ask person to refrigerate all available food eaten during the 72 hours before onset of illness; save or retrieve original containers or packages; sample should be properly identified; hold until health official makes further arrangements. Save any water in refrigerator and trays of ice cubes in freezer; collect was sample from suspect supply in clean jar; put on lid and refrigerate.
* Enter onto complaint log (Form B).
† Enter onto complaint log (Form B) under comments. USE REVERSE SIDE OR ATTACHED SHEET IF MORE SPACE REQUIRED FOR ANY ENTRY

FOODBORNE, WATERBORNE, ENTERIC ILLNESS & COMPLAINT LOG

Form B

COMPLAINT[1]			ILLNESSES			FOOD		WATER[2]			HISTORY OF EXPOSURE[3]	COMMENTS
No.	Date	Type	Onset date	No. ill	Predominant symptom/ sign	Alleged/ Suspected	Where eaten within 72 hrs*	Where ingested within 2 wks*	Where contacted within 2 wks*	Source*		(Specify place, location)

Legend: [1]Type complaint – I = illness; CF = contaminated/adulterated/spoiledfood; UE = unsanitary food establishment; DW = poor quality drinking water; RW = poor quality recreational water; MP = complaint related to media publicity; D = disaster.

[2]Water source – C = community; NC = non-community; O = other; W = well; B = bottled; S/L = stream/lake; V = vended; U = untreated; O = other

[3]History of exposure (specify by name) – DT = domestic travel (out of town but within country); IT = international travel; CC = child care; CI = contact with ill person outside household or visitor to household; C = contact with ill person within household; AN = exposure to ill animal.

*Enter each place or source on separate line under the complaint number.

CASE HISTORY: CLINICAL DATA

Form C1

Name			Source or place of outbreak, if known	Complaint number	Case identification no.
			Address		Phone: Home Work
Age	Sex	Occupation	Place of work	Ethnic group, special dietary habits, immunocompromised or other pertinent personal or health data	

Signs and Symptoms† (Check appropriate signs and symptoms and circle those that occurred first)

INTOXICATIONS (Acute and chronic) | **ENTERIC INFECTIONS** | **GENERALIZED INFECTIONS** | **Other INFECTIONS** | **NEUROLOGICAL ILLNESSES**

INTOXICATIONS (Acute and chronic):
☐ Nausea
☐ Vomiting
☐ Anemia
☐ Bloating
☐ Burning sensation (mouth)
☐ Cyanosis
☐ Dehydration
☐ Excessive salivation
☐ Flushing
☐ Foot/wrist drop
☐ Insomnia
☐ Metallic taste
☐ Pallor
☐ Pigmentation
☐ Prostration
☐ Scaling of skin
☐ Soapy/Salty taste
☐ Thirst
☐ Weight loss
☐ White bands on fingernails
☐ Others (specify)

ENTERIC INFECTIONS:
*☐ Abdominal cramps
*☐ Diarrhea
  ☐ bloody #
  ☐ greasy
  ☐ mucoid
  ☐ watery
  No./day ____
☐ Chills
☐ Constipation
*☐ Fever ____ °C/°F
☐ Tenesmus
____

GENERALIZED INFECTIONS:
☐ Cough
☐ Edema
☐ Headache
☐ Jaundice
☐ Lack of appetite
☐ Malaise
☐ Muscular aching
☐ Perspiration
☐ Stiff neck joints
☐ Swollen lymph nodes
☐ Weakness
☐ Decrease urine output
☐ Pain in back/kidney

Other INFECTIONS:
☐ Ear
☐ Eye
☐ Itching
☐ Mouth
☐ Rash
☐ Skin lesion
☐ Pneumonia
Describe:
____
____
____
____

NEUROLOGICAL ILLNESSES:
☐ Blurred vision
☐ Coma
☐ Delirium
Difficulty in: ☐ speaking
  ☐ swallowing ☐ breathing
☐ Dizziness
☐ Double vision ☐ Irritability
☐ Disorientation/loss of memory
☐ Hot/cold reversal syndrome
☐ Numbness ☐ Paralysis
Pupils ☐ dilated, ☐ fixed,
  or ☐ constricted
☐ Tingling

Other symptoms	Time of onset Date Hour	Incubation period	Duration of illness	Residual symptoms	Fatal Yes ☐ No ☐

Known allergies	Medications taken for illness		Amount	Dates	Medications/inoculations prior to illness

Physician consulted	Address	Phone	Hospital attended	Address	Phone

Contacts with known cases before illness (names)					

Cases in household occurring subsequently (names)			Dates of onset		Child care exposure (place)

Type of specimens obtained 1. 2. 3.	Date collected	Specimen number	Laboratory results		Case ☐ Confirmed ☐ Presumptive ☐ Suspect
			Laboratory method		
			Laboratory where analysis performed		

†Signs and symptoms are listed in columns to suggest classification of the disease; their occurrence is not necessarily limited to the category in which they appear on this form.
*Ask if these symptoms occurred, even if they were not mentioned in the interview.
#Ask whether there was decreased urine output.

CASE HISTORY: FOOD/WATER HISTORY AND COMMON SOURCES

Form C2

☐ Ill   ☐ Well

	Date	Day before illness outbreak Date	Two days before illness Date
Date of illness/outbreak[1]			
Breakfast[2]	Hour	**Breakfast[2]** Hour	**Breakfast[2]** Hour
Place		Place	Place
Item[3]		Item[3]	Item[3]
Companions[4]		Companions[4]	Companions[4]
Lunch[2]	Hour	**Lunch[2]** Hour	**Lunch[2]** Hour
Place		Place	Place
Item[3]		Item[3]	Item[3]
Companions[4]		Companions[4]	Companions[4]
Dinner[2]	Hour	**Dinner[2]** Hour	**Dinner[2]** Hour
Place		Place	Place
Item[3]		Item[3]	Item[3]
Companions[4]		Companions[4]	Companions[4]
Non-meal snacks/water ingested[2]	Hour	**Non-meal snacks/water ingested[2]** Hour	**Non-meal snacks/water ingested[2]** Hour
Place		Place	Place
Item[3]		Item[3]	Item[3]
Companions[4]		Companions[4]	Companions[4]

Item	Time of eating, drinking or contact		Source	Address
History of ingesting suspect food or water or contact with water from suspect source	Date	Hour		

	Date	Persons attending[4]	Ill	Well	Addresses		Phone
Common events or gatherings							

Nonroutine travel past month (international or domestic/ locations)				Water supply[5]	Sewage disposal	Pet/Animals (kind and number of each)

Water contacted during recreation or work in last 2 weeks		Unusual water supplies ingested in last 2 weeks

Investigator	Title	Agency	Date

[1] If ill before all meals eaten, complete column for three days before illness and so indicate to obtain 72-hour history.

[2] If water suspected, number of glasses of water, number of cold beverages made with water, number of beverages with ice ingested per day.

[3] Include all foods, ice, water, and other beverages.

[4] Record names of persons eating same meal and whether or not ill.

[5] Specify C for community, SP for semipublic, U for untreated, B for bottled water, NC = Non-community, S/L = Stream/Lake, W = Well, V = Vended and O = Other.

CASE HISTORIES SUMMARY: CLINICAL DATA

Form D1

Place of outbreak   Dates of outbreak   Complaint number

ID no.	Name of exposed persons whether or not ill	Address	Phone	Sex	Age	III	Time of Ingesting food or water or contacting water		Onset of initial Symptom		Incubation period (differences between time of ingesting/-contact and onset)	Signs and symptoms							Duration (no. days)	Date			Remarks
							Day	Hour	Day	Hour		Nausea	Vomiting	Diarrhea	Abdominal cramps	Fever				Physician seen	Hospitalized	Death	

Investigator   Title   Median IP   Remarks

CASE HISTORIES SUMMARY: WATER/LABORATORY DATA
Form D2

NOTE: Line up with appropriate identification number on Form D1

Water ingested at suspect meal or event	Amount of water ingested			Laboratory tests			Specific comments or additional information about any ill, not ill persons. (Record all information where space does not permit in other sections, such as additional symptoms, physician, and hospital names.)	Case			ID #
	Glasses water per day	Cold beverages made with water per day	Beverages made with ice per day	Specimen	Date collected	Result		Suspect	Presumptive	Confirmed	

Confirmed etiology

Suspect water

Remarks

CLINICAL SPECIMEN COLLECTION REPORT
Form E

Complaint no.	Specimen no.

Place of outbreak	Address	Case I.D. no.	Type of specimen
Patient name	Address		Phone

Reason for collecting specimen
□ Victim of outbreak   □ Person at risk but not ill   □ Handler of suspect food or water
□ Suspected carrier   □ Animal   □ Other (specify)

Physician	Address		Phone

Symptoms: □ Nausea   □ Vomiting   □ Diarrhea   □ Fever   □ Other (specify)

Time of ingesting/ contacting suspect food, meal, or water Day Hour	Time of onset Day Hour	Incubation period	Duration of illness	Medications Type	Amount	Dates

Method of collecting specimen	Method of preservation	Method of shipment

Other Information

Investigator collecting specimen	Title	Agency	Date Hour collected/submitted

Test requested	Presence/Absence	Count/Titer/ Concentration	Definitive type

Comments and interpretations

Laboratory analyst	Lab name & location	Date/Hour received	Date started	Date completed	Etiologic agent as determined by analyst

Water/Ice Sample Collection Report Form F		Complaint No.	Sample No.
Identification of water supply	Location	Sampling Point	Date/Hour Collected
Person in Charge	Phone/e-mail	Description of Sample Including Amount Sample or Filtered	
Method of Sterilizing Containers[1] and/or Collection Utensils[2]		Method of Transportation of Sample	
Shipped □ Refrigerated □ Frozen □ Ambient temperature		Identification marks	Date/Hour Shipped
Estimated Chlorine Contact Time Before Sampling	Chlorine Free ____ Total ____	Temperature of water	Other Field Test Results
Symptoms of victims □ Nausea □ Vomiting □ Abdominal Cramps □ Fever □ Diarrhea □ Conjunctivitis □ Other (specify)			
Time of Ingesting/Contacting Suspect Water Date Hour	Time of onset Date Hour	Incubation period	Duration of illness
Investigator	Title	Agency	Date/Hour

Test Requested	Presence/Absence	Count/Concentration	Serotype
□ *Campylobacter*			
□ *Cryptosporidium*			
□ *E. coli* (specify type)			
□ *Giardia*			
□ *Legionella*			
□ *Salmonella*			
□ *Shigella*			
□ *V. cholerae*			
□ *V. parahaemolyticus*			
□ *Yersinia enterocolitica*			
□ Others (Bacteria, viruses, parasites, toxic chemicals specify)			
□ Heterotrophic Plate Count			
□ Coliphage			
□ Total coliform			
□ Enterococci			
□ E. coli (indicator)			
□ Total culturable viruses			
□ Other (specify)			

Physical properties of Water: (Turbidity)	pH	Chemical Properties of Water		
Comments and Interpretation				
Laboratory Analyst	Agency	Date Received	Started	Completed
Etiologic Agent				

[1]Attach a list of number, sample, and tests desired for other samples collected at the same establishment during the same investigation

[2]Specify only if unusual (such as field) method of sterilizing/sanitizing collection container or utensil or if an unusual method of collecting sample

ILLUSTRATION OF CONTAMINATION FLOW

Form G1 Diagram defective portion of water supply and illustrate source of pollution and their likely entrance into the system. (Specify gradients and pressure differential that altered flow.)

Complaint No.

Investigator	Title	Agency	Date

Scale 1 block =

RECORD REVIEW OF ON-SITE INVESTIGATION AND TEST RESULTS PRIOR TO AND DURING OUTBREAK

(Data from on-site record review and Forms F and G – Form G2)

Complaint No

Heterotrophic Plate Count (plant/distribution)

Location	Date	Results

E. coli Counts/(raw/plant/distribution)

Location	Date	Results

Other Chemical Tests (e.g. ozone)

Specify	Date	Result

Total Coliform Count (raw water)

Location	Date	Results

Other Microbiological Tests (specify)

Type	Date	Results

Physical/Organoleptic Tests (pH, turbidity, UV light)

Specify	Date	Result

Total Coliform Count (finished)

Location	Date	Results

Chlorine Residuals (plant)

Location	Date	Results ☐ Free ☐ Total

Other Examinations

Specify	Date	Result

Total Coliform Count (distribution)

Location	Date	Results

Chlorine Residuals (distribution)

Location	Date	Results ☐ Free ☐ Total

Interpretations

Reviewer _____ Title _____ Date _____

SOURCE AND MODE OF CONTAMINATION OF SURFACE SOURCES[1]
Form G3

Name of surface supply	Location	Person-in-charge		Complaint No
				Phone/e-mail

LAND USE OF WATERSHED	RECENT DATES	TYPE SEWAGE FOR POPULATED AREAS	DISCHARGES INTO SURFACE WATER	TYPES OF ANIMALS IN WATERSHED
☐ Cultivated ☐ Feedlot ☐ Forested ☐ Industrial ☐ Irrigated ☐ Mining ☐ Oil fields ☐ Pasture ☐ Recreation ☐ Thickly ☐ Other settled (describe) _____	☐ Flooding _____ ☐ Drought _____	☐ Primary ☐ Secondary ☐ Oxidation Pond ☐ Untreated/raw ☐ Septic Tanks ☐ Other (describe)	☐ Yes ☐ No ☐ Yes ☐ No ☐ Yes ☐ No ☐ Yes ☐ No ☐ Yes ☐ No ☐ Yes ☐ No	☐ Livestock ☐ Poultry ☐ Aquatic mammals ☐ Waterfowl ☐ Snails ☐ Other (list)

Sewage outfalls or seepage water (give location and distance from water intake or point of use)

Source of pollution (give location and distance from water intake or point of use)
☐ Feedlot ☐ Slaughterhouse ☐ Pasture runoff into surface water

Results of dye test from outlets or seepage to intake or point of use or other means of evaluation of movement of contaminants

Results of any physical/chemical/microbial test of source water (See Form G2)

Factors contributing to surface water pollution/contamination and outbreak*: ☐ Ingestion of untreated water ☐ Pollution of watershed ☐ Dead animal in water ☐ Animals have direct access to water ☐ Use of contaminated water as an alternative source ☐ Overflow of sewage or outfall near water intake ☐ Drought ☐ Flooding ☐ Other (specify)

Investigator	Title	Agency	Date

[1] Note all that apply. Explain source/mode and describe entry in more detail on back or separate attached sheet
*Record on Form L

SOURCE AND MODE OF CONTAMINATION FOR GROUND WATERS[1]
Form G4

Location	Person-in-charge/Owner	Phone/e-mail	Complaint No

TYPE OF GROUND SUPPLY

☐ Well ☐ Spring ☐ Other
(specify)

State source of information:

TYPE OF WELL

☐ Drilled ☐ Bored ☐ Driven
☐ Dug ☐ Step ☐ Other

TYPE OF SOIL AND AQUIFER

☐ Sand   ☐ Clay   ☐ Loam
☐ Peat   ☐ Gravel ☐ Rocky
☐ Limestone ☐ Other (specify)

DEPTH

Static water _____
Well _____

Excreta disposal; in vicinity of well which may have contaminated ground water:

Type: ☐ Community ☐ Community ☐ Leaking sewer ☐ Septic tank ☐ Cesspool/ ☐ Absorption field ☐ Privy ☐ Toxic waste
        primary   secondary   line   seepage pit   disposal

Distance _____

Type: ☐ Stream ☐ Surface water ☐ Animals ☐ Feedlot ☐ Manure piles ☐ Compost ☐ Dump/landfill ☐ Toxic waste
                                                                                                                            storage

Distance _____

Observed faults in construction/maintenance/operation/ of wells/springs/other ground water source:

☐ Casing ☐ Ground casing ☐ Casing not intact ☐ Animal holes around casing ☐ Platform/apron not intact ☐ Pitless adapter faulty

Depth _____ ☐ Open well/spring ☐ Flooding ☐ Casing top below grade ☐ Other (specify)

Type of pump: ☐ Submersible ☐ Jet ☐ Turbine
☐ Reciprocating ☐ Hand ☐ Gravity ☐ Other
                                        (specify)

Source of priming water _____

Disinfection
☐ None
☐ Failure

Chlorine test
☐ Free
☐ Total

Contamination during pumping: ☐ Unsafe water for priming
☐ Leaks in system under vacuum ☐ Well pit flooded
☐ Pump not sealed to platform/top bushing not closed
☐ Other (specify)

Type repairs made _____

Disinfection following repairs: ☐ Yes ☐ No

Date _____

Results of dye test from outlets or seepage to intake or point of use; or other means of evaluation of movement of contaminants

Results of any physical /chemical/microbial test of ground water (give test done, dates, present/absent/count/concentration, as applicable; See Form G2

Factors contributing to ground water contamination and outbreak*
☐ Overflow or seepage of sewage into well/spring ☐ Surface runoff into well/spring ☐ Contamination through limestone or fissured rock
☐ Flooding/heavy rains ☐ Chemical/pesticide contamination ☐ Seepage from abandoned well ☐ Contamination through suction line
☐ Improper well/spring construction ☐ Unsafe water used for priming ☐ Other (specify)

Investigator	Title	Agency	Date

[1] NOTE ALL THAT APPLY. Explain source/mode of contamination and describe entry in more detail on back on separate sheet.

* Record on Form K

DISINFECTION FAILURES THAT ALLOWED SURVIVAL OF PATHOGENS OR TOXIC SUBSTANCES[1]

Form G5a

Name of facility	Location	Person-in-charge	Phone/e-mail	Complaint No

Type of disinfection: ☐ None ☐ Simple chlorination ☐ Super chlorination ☐ Breakpoint chlorination ☐ Ultraviolet ☐ Hypochlorite
☐ Chloramine ☐ Chlorine dioxide ☐ Ozone ☐ Other (specify)

Deficiencies in: ☐ Disinfection equipment ☐ Disinfection operation ☐ Disinfectant contact time   Comments
Interruption in: ☐ Disinfection equipment ☐ Disinfection operation
Dates

Disinfectant demand   Comments

Disinfection tests at plant (give minimum values)   Location:
During investigation   day before   2 days before   last week (date)   last month (date)   Comments
Free

Disinfectant rate applied = disinfectant used per day / flow rate
Disinfectant demand (usage) = disinfectant dosage applied – disinfectant concentration measured downstream

☐ Sudden changes to disinfectant demand,
If yes, date(s):

Sequence	Disinfectant concentration C (mg/L)	Disinfectant contact time T (minutes)	CT_{calc} C x T	pH	Water temperature (°C)	CT 99.9 (from Table I)	$CT_{calc}/CT_{99.9}$
1st							
2nd							
3rd							
4th							
5th							
Sum							

Factors contributing to survival of pathogen or failure of inactivation of toxin during treatment and outbreak[*]
☐ Inadequate prefiltration treatment ☐ Inadequate filtration ☐ Inadequate chemical feeding ☐ No disinfection ☐ Inadequate disinfection
☐ Interruption of disinfection ☐ Other (specify)

Investigator	Title	Agency	Date

[1] Explain treatment failure and describe entry in more detail on back or on separate attached sheet

* Record on Form K

SOURCE OF CONTAMINATION AND TREATMENT FAILURES THAT ALLOWED SURVIVAL OF PATHOGENS OR TOXIC SUBSTANCES[1]

Form G5b

Name of facility	Location	Person-in-charge	Complaint No
			Phone/e-mail

Raw water intake
☐ Excessive pollution in relation to treatment potential ☐ Bypass connection by which raw or partially treated water gets into distribution system   ☐ Nearby uncontrolled pollution ☐ Other (specify)   ☐ Fluoridation feed deficiencies ☐

Sedimentation deficiencies: ☐ No sedimentation before filtration ☐ Turbidity not removed ☐ Tank not cleaned
☐ High proportion of microorganisms remain ☐ Retention time   ☐ Other deficiencies (specify)

Sedimentation rate: Depth of water / Transit time from inlet to outlet = _____

Record review: Coagulant dose   pH   Turbidity   Other tests (specify)
Date/value: /   /
Records show routine monitoring of measurements: ☐ Yes   ☐ No

Filtration performance criteria: ☐ Media loss ☐ Media deterioration ☐ Mud ball formation ☐ Channeling ☐ Surface cracking
☐ Under drain failure ☐ Cross connections ☐ Chemical deficiencies (specify)

Type filtration: ☐ Conventional (rapid) ☐ Direct (rapid) ☐ Pressure ☐ Slow ☐ Bag cartridge ☐ Diatomaceous earth
☐ Other (specify)
                              Frequency of   Deficiencies of   Recycling back-   Source of
                              backwashing: filtration: wash water: backwash water:
                                                                               ☐ Yes ☐No

Average filtered water turbidity: Filter 1   Filter 2   Filter 3   Filter 4   Filter 5   Filter 6   Other filters (list on back)
                                     _____   _____   _____

Combined filter effluent   Clearwell effluent   Plant effluent

Nature of recent illnesses of staff (name of illness or recent symptoms)   Name of employee

Other observations or measurement of treatment plant operations

Factors contributing to survival or failure of inactivation of toxin during treatment of outbreak*
☐ Inadequate prefiltration treatment ☐ Inadequate filtration ☐ Inadequate chemical feeding ☐ No disinfection ☐ Inadequate disinfection
☐ Interruption of disinfection ☐ Other (specify)

Investigator	Title	Agency	Date

[1] Explain treatment failure and describe entry in more detail on back or separate attached sheet

* Record on Form K

SOURCE AND MODES OF CONTAMINATION DURING DISTRIBUTION AND AT POINT-OF-USE[1]
Form G6

Location	Person-in-charge	Phone/e-mail	Complaint No

Type cross connections: ☐ Sewer lines ☐ Waste lines ☐ Fire water supply ☐ Boilers ☐ Carbonated water lines ☐ Cooling water ☐ Hydraulic operations ☐ Other (specify)

Cross connection deficiencies
☐ Deficiency of double check valve arrangement ☐ Defective check valve ☐ Defective other backflow prevention devices ☐ Temporary attachment not detached ☐ Other (specify)

Comments

Backsiphonage detected: ☐ Inlets without air gap ☐ Inlet too close to fixture side/well ☐ Submersed inlet ☐ Hose attachment in vessel ☐ Defective vacuum breakers ☐ Connections to sprinkler system used to spray pesticide or toxic substances ☐ Negative pressure ☐ Other (specify)

Negative pressure occurred due to:
☐ Water shut off due to repairs   ☐ Nearby fires   ☐ Negative pressure on upper floors
Date(s) ___   Date(s) ___   Date(s) ___
Typical repair(s) ___
Disinfection afterwards ☐ Yes ☐ No
☐ Repumping   ☐ Intermittent service   ☐ Other (specify)
Date(s) ___   Date(s) ___   Date(s) ___

Sites sampled
1 ___ 6 ___
2 ___ 7 ___
3 ___ 8 ___
4 ___ 9 ___
5 ___ 10 ___

Recent illness of persons in building	Date(s)	Name of person(s)	Address	Phone/e-mail
Type of illness/major symptoms				

Chlorine residuals in distribution system	Location	Line pressure testing results	Results of other tests (specify test)
Free Total			

Previous month

No. sites where disinfectant residual was measured (a)

No. sites where no disinfectant residual measured, but HPC measured (b)

No. sites where disinfectant residual not detected and HPC not measured (c)

No. sites where disinfectant residual not detected, HPC criteria (e.g. > 500/mL exceeded) (d).

No. sites disinfectant residual not measured, HPC > criteria (e.g. > 500/mL) (e)

$$v = \left(\frac{c + d + e}{a + b + c + d + e} \right) / \left(\frac{a + b}{a + b + c + d + e} \right) \times \frac{100}{100} = \underline{\quad} \% $$
$$\times \frac{100}{} = \underline{\quad} \%$$

Type of storage/transportation facility contaminated: ☐ Community storage tank ☐ Cistern ☐ Transportation tank ☐ Household storage container ☐ Other (specify)

Factors contributing to distribution line contamination and outbreak*
☐ Cross connections and defective backflow prevention devices ☐ Submerged inlet and backsiphonage ☐ Contamination of storage tank, cistern, storage containers ☐ Improper or no disinfection of mains, plumbing or storage facility, transportation container after repairs or new connection ☐ Line pressure loss ☐ Other (specify)

Comments

Investigator	Title	Agency	Date

[1] Explain source/mode of contamination and describe entry in more detail on back or on separate attached sheet
* Record on Form K

CONTAMINATION SOURCE AND SURVIVAL OF PATHOGENS OR TOXIC SUBSTANCES FOR RECREATIONAL WATERS[1]
Form G7

		Complaint No.

Name of pool/lake/spa/spray pad/water course	Location	Person-in-charge	Phone/e-mail

Likely route of infection
☐ Contact ☐ Inhalation ☐ Accidental ingestion

Type of exposure
☐ Swimming pool ☐ River/stream ☐ Lake/pond ☐ Whirlpool ☐ Hot tub
☐ Spray pad ☐ Other (specify)

Type filtration: ☐ Rapid ☐ Pressure ☐ Diatomaceous earth ☐ Other (specify)	Frequency of backwashing	Deficiencies of filtration	Recycling backwash water ☐ Yes ☐ No	Source of backwash water	Turbidity

Type of disinfection
☐ None ☐ Injected hypochlorite ☐ Batch hypochlorite ☐ Other (specify)

Deficiencies in: ☐ Disinfection equipment ☐ Disinfection operation Interruption in: ☐ Disinfection equipment ☐ Disinfection operation Date(s)	Comments about deficiencies or interruption:

Disinfectant residuals

	Day before onset first case	2 days before	3 days before	Last week	Last month (date)	Calculated usage
During investigation						Sudden changes in disinfection usage Date(s)
Free:						
Total:						

Data for CT calculations: (Compare with data on disinfectants and microbes of concern in Table I and J)

Disinfectant residual (C)   Exposure time (T) CT_{calc} (C x T)   Water temperature °C   pH of water $CT_{99.9}$ $CT_{calc}/CT_{99.9}$

Factors contributing to WATER CONTAMINATION AND/OR SURVIVAL (check all that apply)*
☐ Sewage outflows   ☐ No filtration   ☐ Improper pH adjustment   ☐ Diving in water
☐ Flooding/heavy rain   ☐ Inadequate filtration ☐ Snails present   ☐ Wading/swimming/skiing
☐ Underground seepage of sewage ☐ No disinfection   ☐ Puncture injuries or wounds ☐ Other (specify)
☐ Swimming in parasite-infested water ☐ Improper disinfectant ☐ Rough pool well construction ☐ Animals have access to watershed
☐ Uncapped wellhead

Comments			

Investigator	Title	Agency	Date

[1] Explain source of contamination and treatment failure and describe entry in more detail on back or on separate attached sheet

* Record on Form K

CONTAMINATION SOURCE AND SITES OF AMPLIFICATION AND AEROSOLIZATION OF PATHOGENS[1]

Form G8

Name of pool/lake/spa/spray pad/water course	Location	Person-in-charge		Complaint No.
				Phone/e-mail

Device(s) potentially producing aerosols: ☐ Cooling towers ☐ Evaporative condensers ☐ Humidifiers ☐ Water heaters and holding tanks ☐ Shower heads ☐ Decorative fountains ☐ Dusty environment ☐ Ultrasonic mist machines ☐ Irrigation systems ☐ Other (specify) _____

Condition of device: ☐ Dirt/dust observed ☐ Sediment ☐ Slime ☐ Other (Specify) (Explain)

Amplification potential (explain)

Disinfectant used	Concentration	Frequency	Date of last application

Results of air flow or sampling studies (explain)

Factors contributing to contamination, growth and/or survival (check all that apply)*
☐ Aerosols generated ☐ Close contact or air currents carried aerosols ☐ Susceptible persons (e.g. >50 years old, smokers, heavy drinkers, immunosuppressed) ☐ Warm water conducive to growth ☐ Poorly operated/maintained water system ☐ Other (specify)

Water samples collected (number and location)		Comments

Investigator	Title	Agency	Date

[1] Explain source of contamination and treatment failure and describe entry in more detail on back or on separate attached sheet

* Record on Form K

FOOD/BEVERAGE/COMMUNITY ATTACK RATES

Form H1, Vehicle Attack Rate Table[1]

| Place of Outbreak | | Complaint No. | |

Food/Beverages	Number of Persons Who Ate or Drank			Number of Persons Who Did **Not** Eat or Drink			Difference in Percent	Relative Risk	Statistical Significance
	Ill	Well	Total	Attack Rate	Ill	Well	Total	Attack Rate	

[1]For calculation of data when all persons who may have been exposed to the suspect vehicle (but not all persons will have ingested every beverage/food) have been identified and interviewed (retrospective cohort)

COMMUNITY ATTACK RATE

Exposure	Name of Community	Number Ill	Population of Community	Attack Rate	Relative Risk	Statistical Significance
Exposed (at risk)						
Related, unexposed						
Total						

Remarks and interpretation (compare with laboratory results)

Prepared by: _____ Title _____ Date _____

CASE-CONTROL VEHICLE EXPOSURE/DOSAGE
Form H2, EXPOSURE RATES

Place of Outbreak | Complaint No.

Beverages	Ill				Well				Difference in Percent	Odds Ratio	Statistical Significance
	a Ate/drank	c Did Not drink	a+c Total	Percent	b Ate/drank	d Did Not drink	b+d Total	Percent			

Remarks and Interpretations (compare with laboratory reports)

QUANTITY OF WATER INGESTED

Number of glasses ingested per day[1]	Ill	Well	Total	Attack Rate	Relative Risk	Statistical Significance
5 or more						
3-4						
1-2						
0						
Total						

[1] For calculation when a sample of ill and not ill persons may have been exposed to the suspect vehicle and have been identified and interviewed

Number of Beverages Containing Water[2]	Ill	Well	Total	Attack Rate	Relative Risk	Significance Statistical
5 or more						
3-4						
1-2						
0						
Total						

[2] Include all beverages made of water excluding hot beverages (these are to be included in the attack rate).

Prepared by: | Title | Date

LABORATORY RESULTS SUMMARY
Form I

Complaint No.	Outbreak	Date

Cases (Data from Forms C, D, and E)

I.D. Number	Specimen	Organism/Test result	Marker (e.g. serotype, phage type, PFGE)

Water Environmental samples (Data from Form F)

Sample No.	Sample Description	Organism/Chemical recovered	Count	Marker

Interpretation and remarks

Etiologic agent	Vehicle	Source of contamination

Form J1: Chi-sq. analysis can be easily completed using on-line calculators or statistics programs such as Epi-Info. However, to confirm the result or to do the whole thing yourself, here are the steps:

Calculation example: odds ratio and chi-square (χ^2) statistic

	Ill	Well	Totals	
Exposed	18 _a_	_b_ 12	30	**Step 1:** Create a 2x2 table as shown with observed data (O values) and marginal totals
N/exposed	_c_ 3	_d_ 16	19	**Step 2:** Calculate odds ratio:
Totals	21	28	49	(AxD)/(BxC) = (18x16)/(12x3)= 8.0 95% CL: (1.64 < OR < 44.23)[1]

	Ill	Well	Totals	
Exposed	18 (12.857) _a_	_b_ 12 (17.143)	30	**Step 3:** Enter expected (E) numbers for each cell: E = (Row total)x(column total) / (grand total).
N/exposed	_c_ 3 (18.429)	_d_ 16 (10.857)	19	e.g. For cell (a): (30x21)/49 = 12.857 Complete for all cells.
Totals	21	28	49	Note: Any E numbers less than 5? If yes, then stop and go to Fisher's Exact Test [Form J2]

Step 4: Calculate chi-sq. (χ^2) as the sum of $\dfrac{(O-E)^2}{E}$ for all 4 cells

e.g. for cell (a): $\dfrac{(18-12.857)^2}{12.857} = 2.057$

Chi-sq. (all four cells)=2.057 + 1.543 + 3.248 + 2.436 = 9.284

Step 5: Compare your *calculated* chi-sq. value with the *critical* value to determine significance (Table 17):

Table 17 Critical values of the chi-sq. distribution

1. Locate the row showing your table size; 2. Begin at column $P<0.05$ ... Your calculated chi-square value must meet or exceed the critical value to be considered statistically significant at that level. If you fail to meet or exceed the minimal value for $P<0.05$, the result is $P>0.05$, and the relationship is declared "*not significant*".

Table row × column	for $P<0.05$ chi-square must exceed ...	for $P<0.025$ chi-square must exceed ...	for $P<0.01$ chi-square must exceed ...	for $P<0.005$ chi-square must exceed ...	for $P<0.001$ chi-square must exceed ...
for 2×2 tables (1 df)	3.841	5.024	6.635	7.879	10.828
for 2×3 tables (2 df)	5.991	7.378	9.210	10.597	13.816
for 2×4 tables (3 df)	7.815	9.348	11.345	12.838	16.266
3×3 or 5×2 tables (4 df)	9.488	11.143	13.277	14.860	18.467

[1] Odds ratio shown here with confidence limits. This is normally produced by software programs such as Ep-Info. If limits <u>include</u> 1.0 then the relationship *cannot be significant*, regardless of the Chi-sq. analysis.

Step 6: Summarize: *Exposure was related to illness. **Ill persons** were eight times more likely to have been exposed to this factor than non-ill persons. This relationship is statistically significant. The probability of these data occurring by chance alone is less than 0.5%. <u>Reject</u> the null hypothesis of "no association."*

[Odds ratio: 8.0, 95% CL: 1.64<OR<44.23, chi-sq.: 9.28, 1 df, P<0.005]

Form J2: Fisher's exact test can be easily completed using on-line calculators or statistics programs such as Epi-Info. However, to confirm the result or to do the whole thing yourself, here are the steps.

Calculation example: odds ratio and fisher's exact test

	Ill	Well	Totals
Exposed	8 (a)	5 (b)	13
N/exposed	2 (c)	16 (d)	18
Totals	10	21	31

[Odds ratio: 12.8,   95% CL 1.60 < OR < 131.10]

Step 1: If all E numbers in 2x2 table ≥ 5, use Chi-sq. [see Form J1]

Here, cell 'a' has smallest E value at 4.2

Step 2: Calculate the Probability directly...

$$P_1 = \frac{10! \ \times \ 21! \ \times \ 13! \ \times \ 18!}{8! \ \times \ 5! \ \times \ 2! \ \times 16! \ \times \ 31!} = 0.0044$$

	Ill	Well	Totals
Exposed	8̶ 9 (a)	5̶ 4 (b)	13
N/exposed	2̶ 1 (c)	1̶6̶ 17 (d)	18
Totals	10	21	31

Step 3: If we don't yet have a zero in the cells, add +1 to larger of (a)x(d) or(b)x(c) andsubtract-1 from smaller pair. Keep marginal totals fixed.   Recalculate:

$$P_2 = \frac{10! \ \times \ 21! \ \times \ 13! \ \times \ 18!}{9! \ \times \ 4! \ \times \ 1! \ \times \ 17! \ \times \ 31!} = 0.0003$$

	Ill	Well	Totals
Exposed	9̶ 10 (a)	4̶ 3 (b)	13
N/exposed	1̶ 0 (c)	1̶7̶ 18 (d)	18
Totals	10	21	31

Step 4: We still don't yet have a zero in the cells, so again add 1 to (a)x(d) and subtract 1 from (b)x(c). Now a zero appears.   Recalculate one last time: Cancel where possible.   (Note 1! =1 and 0! =1)

$$P_3 = \frac{\cancel{10!} \times \ 21! \ \times \ 13! \ \times \cancel{18!}}{\cancel{10!} \times \ 3! \ \times \ 0! \ \times \cancel{18!} \times 31!} = 0.000006$$

Step 5: The final probability (*P*) is the sum of all probabilities (in this case $P_1 + P_2 + P_3$) or approximately 0.0047.

Step 6: Summarize: *Exposure was related to illness. Ill people were almost 13 times more likely to have been exposed to this factor compared to non-ill people. The relationship is statistically significant. The probability of these data occurring by chance alone is less than 0.5% (<0.005). Reject null hypothesis of "no association".*
[Odds ratio: 12.8, 95% CL: 1.6<OR<131.1, P=0.0047].

Notes:

1. When deciding which cells to increase by +1, always multiply (a)×(d) and compare with (b)×(c). *Increase* each cell of the pair with the higher product and *decrease* each cell of the pair with the smaller product, while keeping all marginal totals unchanged.
2. The final *P* is an "exact" *P* (probability) and may be reported as such (*P* = 0.0047). In this example, it is also <0.005 of course, and can be reported in this way if preferred.
3. The Fisher's test is used when the Chi-Square test is invalid due to any "E" values <5 in a 2×2 table. In all other circumstances, Chi-Sq. is an excellent approximation for the FE test.
4. If original data include a zero in one of the cells, you will calculate only one *P* value. (The O.R. will be reported as "undefinable" but the direction of the effect will be very clear).

5. This P is calculated for a one-tailed FE test. It is adequate for this application. Two-tailed FE test will require further calculation.
6. Should a relationship NOT meet the critical value for significance (that is, $P>0.05$), it is described as *"not statistically significant"*. Note that a relationship may be observed, but this result is telling you that it could have occurred by chance alone more than 5% of time if you were to repeat the analysis. That may still require further investigation, but from a statistical standpoint, it cannot be claimed as a *statistically significant relationship*.

WATERBORNE ILLNESS SUMMARY REPORT

Form K, page 1 of 2

Complaint No. _____

Disease _____

TYPE OF EXPOSURE

☐ Ingestion
☐ Contact
☐ Inhalation

LOCATION OF OUTBREAK

State/Province _____
City or town _____
County _____

DATE OF OUTBREAK
(Date first case became ill)

Mo. _____ Day _____ Yr. _____

NUMBER OF:

	ACTUAL	ESTIMATED
Persons exposed	_____	_____
Persons ill	_____	_____
Hospitalized	_____	_____
Fatalities	_____	_____

HISTORY OF EXPOSED PERSONS

NUMBER OF HISTORIES OBTAINED _____

Enter the percent of persons with the following symptoms

Diarrhea (>3 stools/day) _____
Visible blood in stools _____
Vomiting _____
Nausea _____

Diarrhea (other): No. _____ /definition
Cramps _____
Fever _____
Rash _____

Conjunctivitis _____
Otitis externa _____
Cough _____
Other (specify) _____

INCUBATION PERIOD (hours)
Shortest _____
Longest _____
Median _____

DURATION OF ILLNESS (days)
Shortest _____
Longest _____
Median _____

ATTACH SUMMARY OF CASE HISTORIES (Form D2) ☐

ETIOLOGY OF OUTBREAK (ATTACH FORM I WITH LISTING OF SPECIMENS EXAMINED AND EPIDEMIC CURVE)

	Diagnosis Confirmed	Diagnosis Presumptive	Diagnosis Suspected	COMMENTS:
Agent (if not known enter, "unk")				_____
Pathogen	_____	_____	_____	_____
Chemical	_____	_____	_____	_____
Other	_____	_____	_____	_____

CHARACTERISTICS OF WATER SUPPLY (if illness acquired by ingesting water). ATTACH ALL APPLICABLE PARTS OF FORM G

TYPE OF WATER SUPPLY

☐ Community or municipal
☐ Subdivision
☐ Trailer Park
   Name _____
☐ Noncommunity (with own water supply)
☐ Camp/cabin/recreation area
☐ School
☐ Restaurant
☐ Hotel/motel
☐ Church
☐ Other (specify) _____
☐ Individual household supply
☐ Bottled water (brand) _____
☐ Other (specify) _____
☐ ozone

WATER SOURCE

☐ Well
☐ River/stream
☐ Lake/pond/reservoir
☐ Spring
☐ Other (specify) _____

WATER TREATMENT PROVIDED
(check all that apply)

☐ No treatment
☐ Disinfection
☐ Chlorine
☐ Chlorine and ammonia (chloramines)
☐ Ozone
☐ Corrosion inhibitor
☐ Unknown
☐ Other (specify) _____

WATER TREATMENT PROVIDED (continued)

☐ Coagulation and/or Flocculation
☐ Settling (sedimentation)
☐ Filtration at purification plant
   (do not include home filters)
   ☐ Rapid sand
   ☐ Slow sand
   ☐ Diatomaceous earth
   ☐ Other (specify) _____
☐ Unknown
☐ Other (specify) _____
☐ Unknown

COMMENTS: _____

WATERBORNE ILLNESS SUMMARY REPORT (continued)	Complaint No.	Disease
Form K, part 2 of 2		

VEHICLE

CONFIRMATION BY (check all that apply)   COMMENTS: _____
☐ Laboratory   _____
☐ Attack rate table and high confidence   _____
☐ Other (specify) _____

ATTACH ATTACK RATE TABLE (form H1): ATTACH LISTING OF WATER SAMPLES EXAMINED AND FINDINGS (Form I)

RECREATION EXPOSURE (If illness acquired by contacting water or ingesting recreation water during activities)

Route of entry
☐ Intentional ingestion   ☐ Contact   ☐ Swimming pool   ☐ River/stream   ☐ Lake/pond
☐ Accidental ingestion   ☐ Inhalation   ☐ Hot tub   ☐ Whirlpool   ☐ Other (specify)

ATTACH ALL APPLICABLE PARTS OF FORM F AND DESCRIBE SETTING

FACTORS CONTRIBUTING TO WATER CONTAMINATION AND/OR SURVIVAL (check all that apply)

AT SOURCE
☐ Overflow of sewage   ☐ Use of a backup source of water by a water utility   ☐ Other (specify) _____
☐ Flooding heavy/rains   ☐ Improper construction or location of well or spring   ☐ Unknown
☐ Underground seepage of sewage ☐ Contamination through creviced limestone or fissured rock

AT TREATMENT PLANT
☐ No disinfection   ☐ No filtration   ☐ Other (specify) _____
☐ Temporary interruption of disinfection ☐ Inadequate filtration   ☐ Unknown
☐ Chronically inadequate disinfection   ☐ Deficiencies in other treatment processes

IN DISTRIBUTION
☐ Cross contamination   ☐ Contamination of mains during construction or repair   ☐ Other (specify) _____
☐ Back siphonage   ☐ Contamination of storage facility   ☐ Unknown

OTHER REASONS FOR CONTAMINATION OR SURVIVAL OF ETIOLOGICAL AGENT IN WATER
(include recreational exposure)

ATTACH NARRATIVE REPORT (include unusual aspects of outbreak or investigation not covered above and recommendations given for prevention)
REMARKS:

NAME OF REPORTING AGENCY	PERSON COMPLETING FORM (Please print)	DATE INVESTIGATION INITIATED	
ADDRESS	TITLE	PHONE/E-MAIL/FAX	DATE OF REPORT

Index

A

Aerosolization, 47, 54
Air conditioning, 33, 47, 54
Alert, 2, 10–15, 34, 35, 44, 65, 85, 125
Assistance, 28, 49, 86
Attack rates, 64–66, 72, 74, 76, 77

B

Blood, 20, 21, 90, 92, 94–99, 101–114
Blue green algae, 49, 55, 106, 110

C

Cadmium, 92, 107, 119
Campylobacter, 14, 94, 118
Campylobacter jejuni, 94, 118
Case definition, 22, 23, 57, 66
Case histories, 47
Case-control, 9, 64, 66, 72
Chi-square, 65, 68–70, 72, 76, 77, 145
Clinical specimens, 9, 10, 14, 18, 20–22, 29,
    78, 84, 118, 119
Clostridium perfringens, 81–83
Coliforms, 4, 12, 78, 80–84
Coliphage, 83
Collect water samples, 47, 75, 80
Complaint, 8–15, 21, 28
Contact time, 25, 31, 38, 40–43
Contamination, 1–4, 6, 8, 9, 24–26, 28–56, 64,
    72, 76–85, 93, 96, 101–104, 122
Contributory factors, 26, 30, 85
Control, 2, 4, 7, 15, 18, 25, 26, 34, 45, 47, 54,
    57, 62, 65, 70, 84, 85, 90, 113, 121

Cryptosporidium, 5, 15, 25, 40, 49, 54,
    85, 99, 120
Cyclospora cayetanensis, 100, 120

D

Diagnosis, 14–23
Dose related, 75

E

Emergency, 2–4, 6–9, 26, 28
Epidemic curve, 35, 57, 58, 60, 61, 84
Epidemiologic associations, 8, 15
Equipment, 7, 14, 30, 35, 38, 40, 42, 44–47,
    49, 54, 90–91, 93, 107
Escherichia coli (indicator), 54, 78
Escherichia coli O157:H7, 11, 12, 82
Exposure rate, 62–63, 66, 72, 73, 92

F

Fecal coliform, 78, 82, 83
Fecal streptococci, 82
Feces, 10, 19, 20, 31, 81–83, 93–101,
    103–105, 109, 113
Fisher's exact test (FET), 65, 68, 70–72, 75,
    147
Forms, 7, 14–16, 18, 21, 29, 30, 34, 36, 38, 41,
    45, 50, 51, 59, 84, 90, 123

G

Ground water, 29, 31, 34–37, 92, 104–106, 117

© International Association for Food Protection 2016
Food Protection International Association, *Procedures to Investigate
Waterborne Illness*, DOI 10.1007/978-3-319-26027-3

Printed in the United States
By Bookmasters